T0296718

The Toyota Kaizen Continuum

A Practical Guide
to Implementing Lean

The Toyota Kaizen Continuum

A Practical Guide
to Implementing Lean

John Stewart

CRC Press
Taylor & Francis Group
Boca Raton London New York

CRC Press is an imprint of the
Taylor & Francis Group, an **informa** business

A PRODUCTIVITY PRESS BOOK

First published 2012 by CRC Press

Published 2019 by CRC Press
Taylor & Francis Group
6000 Broken Sound Parkway NW, Suite 300
Boca Raton, FL 33487-2742

© 2012 by Taylor & Francis Group, LLC
CRC Press is an imprint of the Taylor & Francis Group, an informa business

No claim to original U.S. Government works

ISBN-13: 978-1-4398-4604-9 (pbk)

Library of Congress Cataloging-in-Publication Data

Stewart, John.
  The Toyota Kaizen continuum : a practical guide to implementing lean / John Stewart.
    p. cm.
  Includes bibliographical references and index.
  ISBN 978-1-4398-4604-9 (pbk. : alk. paper)
  1. Industrial management. 2. Lean manufacturing. 3. Industrial productivity. 4. Toyota Shatai Kabushiki Kaisha. I. Title.

HD30.3.S74 2012
658.4'013--dc23   2011033623

Visit the Taylor & Francis Web site at
http://www.taylorandfrancis.com

and the CRC Press Web site at
http://www.crcpress.com

Contents

Preface

When I began my career with Toyota in 1989, I had no idea of the path that I had undertaken. I started as one of the first production line employees. Toyota called us *team members*. When I left Toyota in March 2007, I was responsible for Toyota's largest European division. Over the course of my eighteen years with Toyota, I had the opportunity to learn many things that have defined me as an operator of companies today.

Today I work with a private equity firm where I am responsible for managing the operations of a diverse portfolio of businesses. As an investor, our job is to work with the management team of the companies to create value. The value we create is how we create a return for our investors. In both environments, I have learned many things and have successfully applied them to create value. However, during this transition, I have identified some of the barriers that companies that lack the resources and the structure of a company like Toyota face when trying to follow the principles of the Toyota Production System (TPS).

There are too many books that are purely academic exercises that have no real substance and no real merit for the majority of businesses trying to create value in these difficult times. I find it humorous that some have successfully regurgitated Toyota philosophies by defining a system based on a utopian operational environment and slapping a badge of authenticity on the cover. Some of the more recent publications seem to be more of a publicity stunt for a Japanese company that wants to control the image of how it is perceived. Even if this is the case, that is really of no concern to me. My concern is that there are a lot of business leaders who are looking to drive real operational value through their organization. I have written this book to provide real insight into how to use the TPS to drive operational value in any organization.

I don't fault the authors of the above books for their failure to provide real insight into the system. The majority of misconceptions concerning the TPS originate from the sheer number of tools Toyota has developed for implementation. Many organizations are successful at implementing some of the tools of TPS; however, without an adequate roadmap, they often find themselves wandering without a real vision of what needs to be accomplished.

When I was with Toyota I spent over eighty-eight weeks in Japan at the Toyota factories learning from the modern day TPS masters. These are people who will never write a book, because they are too busy actually implementing the system. I have written this book in order to provide the reader with the insight from these masters.

I not only had the opportunity to learn from the masters, I was also responsible for training others on a global basis on the TPS. During the course of my career of training people about its various aspects, I have come to find that there are three types of people who are teaching the TPS:

1. Self-Proclaimed Master—Those who can teach but have never done it themselves
2. Master—Those who can master the skills but cannot teach others
3. Master Creator—Those who have mastered the skills by doing and can teach others

Even inside of Toyota, there were those who would teach but could not do it for themselves. This was one of the reasons that I had such a loyal following in Toyota. I was able to successfully teach others because of the many successes that I had by implementing these principles.

During my career, I have talked with a wide cross-section of people, manufacturers and non-manufacturers alike, who take the Toyota tour with the hope of gaining a more complete understanding of the TPS. Most people recognized the terms *just in time* and *jidoka* as principles of TPS, yet they would look at tools such as kanban, andon, and others as TPS principles as well. Almost without fail, they came in with the preconceived notion that TPS is a fixed system, that there is standardized work from start to finish on what to do, the equipment to do it with, and the manner in which it is implemented. Actually this is far from the case. Even inside Toyota!

In truth, the only inflexible aspects of TPS are the principles of just in time and built-in quality. Everything else is simply a tool for helping your organization to do whatever it is that you do the best way possible.

Just in time and jidoka are the driving principles behind *everything* that Toyota does. All the tools mentioned before are valid, but they exist solely to facilitate flow and quality. If you strive to understand the core principles, you will gain a better understanding of the outlying principles as well. If the tools are used without the core principles behind them, TPS ceases to be a system and becomes a short-term fad.

The real questions we must consider are what is the best way to ensure just-in-time delivery, and what is the best method to build quality into the process. Not every solution is the right solution based on the current condition, so by asking ourselves what is the best way to achieve the ideal condition, we challenge ourselves to get better. To think that TPS is an inflexible system and once it is in place the money will flow and problems will disappear will only cause giant economic headaches. Across the organization, Toyota makes *thousands* of changes a day based on the feedback it receives from its operators and workers.

TPS is a system that searches for the best method to get those thousands of changes, the small ideas and innovations that, across the board, are expected of everyone, from the top floor to the shop floor.

The goal of this book is to give you practical examples of how to utilize the principles of the TPS to drive value in your organization. This book is meant for people who leave work every day with their hands dirty and with a sense of pride in what has been accomplished by their efforts. Enjoy.

The Author

John Stewart is the operating partner for Monomoy Capital Partners, a New York-based private equity fund. At Monomoy, John is responsible for portfolio management. He has worked with numerous businesses of various sizes to increase their bottom line results. Prior to joining Monomoy, John worked for Toyota Motor Corporation for eighteen years. He started with Toyota on the shop floor and worked his way through each level of the organization until he found himself responsible for Toyota's largest European manufacturing division. John was selected by *Automotive News* as one of the top young professionals in the automotive industry in 2007.

John is married to his wonderful wife, Leslie, and they have four children: John II, Sarah, Andrew, and Matthew.

Chapter 1

Introduction

1.1 Don't Believe Everything That You Read in a Book

Today there are more sources of information than ever before that revolve around Toyota and its legendary production system. No matter how it is labeled, the Toyota Production System (TPS) is simply a logical, common-sense approach to manufacturing. Unfortunately, most of the available information only concerns the theory of application and offers no valuable insight into the practical implementation of TPS. This leads the general public to the dangerous assumption that Toyota's manufacturing operations are a utopian environment. The people who work in Toyota would be the first to say that this is far from reality.

Having worked for Toyota for eighteen years, I can truly say that I have nothing but admiration for all of the people who I worked with through those years. The opportunity to work for a company that started as a small import car manufacturer with little-known models (who knew what a Camry was in 1987?) and grew to become the largest manufacturer of automobiles in the world has given me unique insights into the application of the TPS in various environments.

The truly fascinating aspect about all of the things that have been written about Toyota is that Toyota would never say these things about itself; this goes against the true culture of modesty at Toyota. I remember one occasion, when I was working at the Toyota facility in Georgetown, Kentucky, and we had been invited to visit one of our suppliers to review their improvement activities. I was traveling with one of Toyota's renowned experts on the TPS who had the well-deserved reputation as a knowledgeable and stern sensei when it came to adhering to the principles of TPS. He had reprimanded me

Figure 1.1 Disorganized Plant.

on many occasions for what many would consider trivial issues at our facility in Georgetown. Given his proven reputation as a hardass, I was curious to see his response to one of our supplier's facilities where they were still in a stage of infancy when it came to implementation of the TPS.

As we arrived at the plant, the first thing I saw were old pallets stacked haphazardly against the side of the factory, followed by a graveyard of obsolete equipment quietly rusting in an adjacent field. As I turned into the parking lot of the facility, I thought to myself, "The management team of this facility had no idea what they were in for." For some reason all I could think of was a time when my sensei had been touring my facility and had noticed the label on the back of a parts rack, known in Toyota as a flowrack label, that had a trivial discrepancy with the standard.

My sensei had lectured for what seemed like hours on the process and methodology of the kanban and how the flowrack was only to hold no more than two hours worth of stock and why two hours and not two hours and one minute, etc. For a facility in such a state of disarray, I was expecting the reprimand for the plant manager of the supplier's facility to be of epic proportions.

We were greeted by the president of the company and the plant manager in a conference room. As we exchanged pleasantries, they shared with us their understanding of the TPS and what they considered to be their operating philosophy. We were scheduled to go on a plant tour after lunch, but my curiosity got the best of me, and I asked if we could go to the shop floor first; the plant manager gladly agreed to my suggestion.

The degraded exterior of the plant was, unfortunately, an accurate indicator of the interior. I was beginning to feel bad about the criticism that I knew was coming. I just hoped that I could somehow elude the onslaught. After years of experience at Toyota, I had thickened my skin to the point where criticism was taken professionally instead of personally. At Toyota, everything was viewed from the standpoint that there was always an opportunity to improve. Even when we reached a target, we would be criticized that the target had been too low, etc.

After visiting the shop floor, it was obvious to me that this facility and the management team did not have this same frame of reference. While the plant manager was busying himself showing us the operations and the plan for improving the operations, I studied my trainer's body language, looking for signs of the reproach to come.

To my amazement, we finished the plant tour without incident! Not one criticism from my sensei. We returned to the board room and had lunch with the president, plant manager, and the rest of the management team. The president asked my sensei what he thought about the facility and its current operational initiatives, and where he thought improvement was needed. I was wearing my best poker face and thought to myself, "Hold on, here it comes." I watched as my sensei stood up and politely thanked them for having us in their facility. He then spent the next thirty minutes telling them all of the *good* things he had seen on the shop floor. Hoping that my face did not reveal the shock that I felt on the inside, I listened intently to his praise for what he termed best practices. When he had finished his praise, he told them that they may realize additional opportunities by emphasizing standardization and workplace organization. I sat in my chair momentarily stunned and thought, "That's it? You have got to be kidding me, this place sucks!" We exchanged our goodbyes and set a date to return in three months time.

As we made our way back to the plant, at first we rode together in silence. After finally trying to come up with the right words, I asked my sensei why he did not take the opportunity to point out all the areas in the operation where there were serious concerns. I reminded him of how he would always find the smallest errors at the plant in Kentucky and deliver a browbeating lecture to me and my team. It was then that he revealed something to me that to this day I have found very valuable; he reminded me that Toyota had been working for over fifty years to implement TPS, and although we did many things correctly, we still had a long way to go. Since we still had so much opportunity and room for improvement ourselves, we should always be *humble* when working with people trying to implement the TPS. In regard to the company we had just visited, the condition of the facility was obvious. Had our goal simply been to measure them based on the condition of our facility, then we could have spent hours pointing out all of the concepts that had been misunderstood and the obvious areas of concern. However, the goal of our visit was to encourage them to continue looking at their operation with a critical eye, looking for opportunities of improvement; therefore, it was much more beneficial for us to develop a relationship of trust and make it our duty to *teach* them to see the things that we had observed and were obvious to our trained eyes. The only real way that they were to improve their factory would be for them to see what we saw and take action based on their own understanding.

My sensei explained that since the president and the plant manager had visited our facility earlier, they understood what a finely tuned operation looked like. He even believed that they were ready for us to tell them a lot of negative things about their operation. Therefore, what benefit would that have had for the plant management and in the long term for our supplier? By taking the opportunity to point out everything that was seen as positive about their efforts, my sensei had disarmed them and therefore the management team was more open to our suggestions. By utilizing this method, my sensei had been able to focus their efforts on the aspects that would benefit them the most. He explained to me that had he chosen to be stern and point out everything that was wrong, it was very possible that they would not have asked us to return, and this could have possibly discouraged their improvement process. This not only would have been bad for them and their employees, it would not have benefited us at our facility in Georgetown either.

As I listened to the words of my sensei, I was reminded of a lesson that I was taught as a child; always show respect while in another person's home, as you are not only representing yourself, but your family as well. This story of the supplier's efforts to implement TPS illustrates the true essence of Toyota culture; it is built upon modesty, not arrogance. Once arrogance enters the system, complacency is not far behind. Many of the books concerning Toyota on the market today have not done justice to the philosophy of modesty that is so important to the culture of Toyota. This is something that Toyota themselves have recently been learning the hard way. With all of the growth that Toyota has seen over the last ten years, there was a big push to bring in executives from other auto manufacturers, mainly the U.S. three (GM, Ford, and Chrysler). Such an influx of senior leaders in the Toyota organization in North America has not allowed the basic principles of Toyota to be thoroughly understood; as a result, modesty has given way to arrogance.

Another fallacy found in many current books is that Toyota is the picture of perfection. Most of the material does a wonderful job of telling the story of how things should operate inside a facility that embraces the essential philosophy of the TPS. There is little reference to the problems caused by implementing the TPS. Problems exist for every organization that has ever tried to implement lean manufacturing concepts, even inside Toyota facilities.

During the years of Toyota's growth, there were numerous occasions when things did not go as planned. Implementing TPS cannot only be costly, but it can also cause significant problems and pose a severe risk to the stability of the operation if not managed correctly. Some authors insinuate that the TPS is the perfect way to manufacture products; this is just not the case. The *search* for the perfect way to manufacture products is the TPS.

Take a mountain climber, for instance. Mountain climbers have to prepare themselves for months and sometimes even years before setting out to climb a mountain. They study all facets of the mountain, the terrain, the geology, the weather, and they even spend time acclimating their bodies to the conditions of the mountain. If the only purpose of a mountain climber is to get to the top of the mountain, there are many more efficient ways to get to the top of a mountain than to just climb up the mountain. However, the accomplishment for the climber does not come from the sole act of reaching the top of the mountain itself; it comes from the complete journey to get there. Climbers often climb the same mountain multiple times. When, at the end of their climbing career, they are telling stories to their friends about the climbing experiences, they may focus not only on the climbs that were successful, but on the failures as well. For a mountain climber the ultimate success may come from reaching the elusive peak of the mountain. Often, however, the most rewarding part of the journey is a point on the mountain where it did not look as though they would be successful. It was at this moment that a decision had to be made based on the progress that had been made, their physical condition, and the resources remaining. This same analogy is true for those who have had the experience of implementing the TPS. Some refer to this process as their lean journey.

A true student of TPS is only happy when he or she is placed in a nearly impossible situation with little or no resources and has to find the way. This is the indispensable attitude that is lacking in those managers and executives who only stand on the sidelines and cheer versus those who actually prepare themselves and participate. This is one of the challenges facing Toyota today. Newly hired executives in the United States who do not have the benefit of having grown up through Toyota's system lack insight into the basic foundational principles of the TPS. Toyota's ability to properly train senior managers going forward will define whether Toyota will be able to work through the current problems being experienced in the Toyota of today in order for the Toyota of the future to be more representative of the Toyota of yesterday.

Just googling "Toyota books" will return over one million hits in a fraction of a second. I actually enjoy reading some of the various books and articles that abound on Toyota and the TPS. I find it amazing that someone can tour a Toyota facility for a few days and author a book that restates everything that is already known, without providing any real insight into the actual process of implementing the TPS. Based on the fact that Toyota's system is a process-driven system, this is counterintuitive. These materials are disappointing from a content standpoint, as they tend to leave the reader with a void. Unfortunately, most often the void is the lack of any real substance that will lead the reader toward a further understanding of how to put any of the concepts into action.

How can you learn to drive a car from someone who has never driven a car? Although this sounds ridiculous, this is exactly what is happening at many universities, manufacturing facilities, health care providers, and offices across the country today. People who have spent time writing books glorifying Toyota in every way possible leave a path of dissatisfied executives who have tried to follow the principles laid out as "Toyota principles" only to end up with a very un-Toyota result. My goal for writing this book is to provide readers with an understanding of the topics that can be readily utilized to take immediate action in their respective organizations.

1.2 ABC's of TPS

During my tenure at Toyota, many people would request to visit one of our facilities. Whenever we had guests at Georgetown, I would be part of the group that met with the visitors to try to explain what they had seen during their visit. Generally, people would visit the Georgetown facility to gain a better understanding of the company and see how the production system was applied for *everyday use*. Many times the visitors would actually be competitors who would come for the plant tour looking for the "secret" of Toyota. While showing them the facility, I would explain the philosophy and purpose of the TPS, and there would be an expectant look in their eyes, as if I were about to produce some magic that they could take back and immediately implement into their own manufacturing process. That look would gradually fade, only to be replaced with an impassive face and suspicious eyes; they always thought I was holding out. The problem for them was that the solution they were searching for was too simple for them to realize that it was a solution at all.

Without fail, when the opportunity to ask questions arrived, they would start to ask very specific questions about *this* specific process or *that* specific piece of equipment. They were searching for something, even though they did not know what they were searching for. They believed that they would know it when they had seen it or when they had heard the correct answer. One time I actually had someone say, "Now that you've shown us everything on the tour, why not let us wander around on our own?" I was a little bit taken aback by the request, given the fact that the facility in Georgetown is over seven million square feet of manufacturing and offices with over seven thousand employees. I tried not to be rude and asked the visitor if it was common practice at his facility to allow visitors to wander around freely inside his facility. Of course, he said no. I explained to him that I was trying to be completely open with him about everything that we did and I was not hiding anything. I asked him what he was really seeking from the visit that he had not been able to ascertain from what I had already presented. He said that he *knew* that we *had* to be hiding something from him. He said that all we had shown were basic manufacturing principles and processes. He said that there had to be some piece of equipment that gave Toyota a competitive edge, and all he could see was very simple equipment that could be found at almost any auto manufacturer. I told our visitor that I had attempted to be completely open and that I would be happy to show him anything that he would like to see and to answer any questions. However, if he wanted to understand the secret of Toyota, then I would explain that to him as well. I explained that the secret to the TPS is not a piece of equipment or a specific method, and if he really wanted to understand the secret to Toyota, all he had to learn was his ABC's. He gave me a confused look and asked me to explain. This is what I explained to him:

Figure 1.2 ABC's.

As schoolchildren, we were taught to read by following a specific process, the same process that is followed to this day. We did not simply pick up a book and start reading; there were a series of steps that we followed. Before we could learn to read, we had to be taught how to make basic sounds; I very clearly remember being confused about all of the different sounds that each letter had. Before we could truly learn those sounds, we had to be taught the alphabet. Since the teacher knew that learning our ABC's would help us to understand the different sounds that would eventually enable us to put those sounds together into words and develop the foundation that we would need to read, the teacher spent much of her time making sure we understood all of the letters and their proper order.

Stephen Covey teaches in *The 7 Habits of Highly Effective People** that you have to begin with the end in mind. This concept, although simple, is also quite profound. Unfortunately, many of us want to not only begin with the end in mind, we also want to finish with the end in mind as well, and the quicker the better. We have a tendency to search for the easy way, or a short cut, and although this is not always a bad thing, we have to understand when it is appropriate. When we learn our ABC's, we cannot just be taught the A and the Z. No, we are taught that first there is A, then B, then

---

* Covey, Stephen R, 1989. *The Seven Habits of Highly Effective People.* Fireside. New York.

Current State   Ideal State

ABCDEFGHIJKLMNOPQRSTUVWXYZ

Figure 1.3   ABC's of TPS.

C, and so on, until we reach Z. Our teachers and parents even taught us a song using the ABC's to help us remember the order. I bet that you can remember that song even now. It is this simple yet solid foundation that allows us to learn the difference between a consonant and a vowel and the individual sounds of each letter. It is only after we have this complete understanding of our ABC's that we are able to combine letters into words and words into sentences.

So what does this have to do with TPS? Simply put, understanding the TPS is the same concept as the ABC's. Even though we know that we are at some point along the path from A to Z—some would call this the lean journey—we know that the destination of our journey is to end at Z.

In Figure 1.3, J is the current state and Z represents the ideal state. Even though we know that Z is much better than J, it is not possible to get to Z from J without moving next to K. It is only through the progression from J to K that we will gain the knowledge and understanding necessary to master K and then one day move on to L. The real improvement is not realized getting to Z; it is the process of getting to Z that has the real value.

1.3 The Kaizen Continuum

One way to look at the progressive process of the TPS is to look at the continuous improvement cycle. I refer to this as the kaizen continuum (Figure 1.4). Similar to the ABC model, where we are is the current situation and once we understand where we need to be, the ideal situation, we can identify the steps necessary to get there. I like this illustration because it makes some basic concepts clear. First we have to assess the situation and understand where we are, and based on where we are we then need to identify the ideal situation. Even though we may know where we need to end up, we cannot simply move from current to ideal, in the same way that we could not move from J to Z in the preceding example. This shows why this process is called the continuous improvement cycle; it is a cycle that must be advanced one step at a time. For us to move from the current situation to the next step, we have to standardize the current

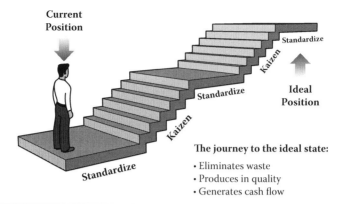

Figure 1.4 Kaizen Continuum.

situation. Once the current situation has been standardized, then we can understand what the next step is and develop a kaizen plan to move to the next step. Of course it would be great if we could move directly to the ideal state, but that is the fascinating aspect of the TPS. Once you start along the journey, you are always measuring yourself to the ideal condition and the closer you get to the final destination, the more you realize how far you actually are from achieving the ideal condition. In the following chapters, we discuss how to begin, maintain, and sustain the cycle in your daily operations.

Chapter 2

Foundational Elements of the Toyota Production System (TPS)

2.1 An Overview of the Toyota Production System (TPS)

In the automobile industry, the name Toyota carries a reputation of unsurpassed manufacturing efficiency. With their almost total domination of the auto manufacturing industry for the last twenty-five years, Toyota has built a foundation that has sustained them as a true manufacturing giant. That foundation is the Toyota Production System (TPS).

Over fifty years ago, Taiichi Ohno devoted his life's work to developing what would become known as the most versatile and productive manufacturing system in the world. Based on a commonsense approach to manufacturing, it became a system synonymous with quality, flexibility, and profitability. Over the years there have been countless books, consultants, and self-described "gurus" who have claimed to have some secret knowledge pertaining to the TPS. I have personally encountered companies that have attempted to follow the direction of some of these individuals only to see them spend millions of dollars to end up really confused. As discussed in the previous chapter, the real secret to the TPS is that there really is no secret. The TPS is a systematic process for improving operations in a company, enabling the company to lower costs. One could say that the "secret" is the system, but Toyota has never gone to great lengths to conceal it; therefore I would not refer to it as a "secret."

As you will discover in this book, the TPS is a systematic approach that when applied to the operations of a company, will drive down operating costs. Although the most common point of reference for TPS is the production of automobiles, TPS has been implemented in the production of a vast array of products, goods, and services; from construction to dentistry, the influence of TPS continues to grow. Known in the Western world as lean manufacturing, the terms are interchangeable once the basic concepts are understood. My concern is that many people use lean manufacturing as a catch-all for all improvement activities. I have yet to meet a plant manager who has not told me, "We have done the whole lean thing." This comment itself gives me insight into their understanding, or lack of understanding, of the fundamental principles of lean manufacturing.

2.2 Toyota's Recent Turmoil

As has been witnessed recently in the news concerning Toyota, an organization is only as good as its people. Although Toyota has made some missteps in how it has handled some situations, this is a reflection not so much on the TPS as on the individual leaders in the company making these decisions. As mentioned earlier, one of Toyota's key factors for success is to remain humble and not to become arrogant. Over the course of the last ten years, Toyota's leaders have focused the company to become the world's largest automaker, even going so far as to pronounce this goal of growth as their 2010 vision for the company in 2004. Like any organization that has an operating system, the system is only applicable as long as the system is understood and followed by the people in the organization. As Toyota began to focus on the 2010 vision to become the largest global automaker, the resources of the organization became stretched, and a key decision was made that has turned out to be a massive mistake. This mistake was the plan that was developed to fill the void in the leadership positions while the company was busy expanding at an exponential rate. Up until the year 2000, Toyota had filled most internal leadership positions with candidates who had been through a rigorous internal training and development program. As the company began to expand in North America and China, the strain on the organization's resources was too much for the company to bear and the company decided to look outside of their internal succession models for external candidates. The lack of internal candidates was due mainly to Toyota's severe standard for plant management that requires that the president for regional manufacturing locations be older than fifty.

Based on these constraints, Toyota began bringing in senior management from other auto manufacturers, mainly Ford, GM, and Chrysler. Some of these hires were good for the company; they brought in some fresh management perspectives and were quick to learn the methodology of the production system. Others, unfortunately, were very limited in their knowledge of Toyota's system and culture and quickly started to manage the organization based on the principles of their former companies. This divergence by Toyota from the system, by hiring managers and leaders who did not have the knowledge of the manufacturing system, is what currently has the future of the company in jeopardy. The system itself is not the problem; the problem is the people managing the system. This is something that Toyota would see for itself, if they could clear away their own arrogance. Only time will tell.

2.3 A History of the Toyota Production System (TPS)

One question that I am often asked when introducing people to the concepts of the TPS is "Why is the understanding of the TPS so important to the world of manufacturing?" The answer may be surprising to some and obvious to others. If we look at the manufacturing industry today, we can see that the impact the TPS has made across the industry is nothing short of astounding. Unfortunately, the impact has been marginalized by the leaders of industry who refer to lean manufacturing, the TPS, and other Japanese manufacturing systems, as a catch-all for continuous improvement.

Fundamentally, the most ignored and overlooked aspect of any successful organization is the management and leadership of the organization. Many books have been written that compare Japanese and Western management styles. Although understanding the principles of different management styles is important, ask any human resource professional and he or she will tell you that there is not one management style that works for all employees. The traditional Western manufacturing methodology that was used thirty years ago no longer has application in today's corporate environment. This relevance has less to do with the influence from Japanese manufacturing techniques than it does with the evolution of the Western worker. An interesting aspect of the TPS is that it is an all-inclusive system for operations and management.

This fact has puzzled many in the automobile manufacturing industry for years. It is ironic that Ford has poured millions of dollars into copying the TPS at many of its production facilities, while Toyota gives Henry Ford a lot of credit for inspiring Taiichi Ohno, the founder of the TPS. It was Henry Ford's advancement in manufacturing techniques, specifically the invention of the automated assembly line, that can be seen in almost every manufacturing facility in the modern world. Henry Ford was also a visionary when it came to the elimination of waste. It is a well-known story that Ford had the pallets that were used for transporting engines to the facility utilized as floor boards in his early vehicles. The pure genius of Henry Ford has much more to do with the advancement of manufacturing from a "eureka moment" perspective than anything done in the last hundred years of manufacturing. Taiichi Ohno only took the basic mass production concepts being used at the time and adapted his commonsense approach to all things manufacturing to develop what is now known as the TPS.

All of the mainstream auto manufacturers today have production systems that are based on the TPS. Whether it is called the Ford Production System, the Nissan Production System, or other, the concepts are all similar. The companies that have been the most successful are the ones that realize that the best way to develop their own production system is to adapt the philosophies and fundamental principles of the system to the culture and circumstances of the organization. For every company to seek to operate exactly the same as Toyota goes completely against the essence of the TPS. This is even true inside of Toyota itself. If there was only one right answer for manufacturing vehicles, then we would expect every Toyota production facility to be exactly the same. However, that is not the case. Of course there are similarities, but on close examination there are many differences.

In today's global economy, it is very gullible for an organization to think that a production facility in the United States is going to operate the exact same way as a production facility in China. The culture of the workforce is as much a factor for developing a successful manufacturing system as the manufacturing system itself. In the 1990s, General Motors started an aggressive campaign to implement the TPS at their facilities in Europe. Even though they heavily recruited Toyota employees to direct the project, the project had mixed results. In the plants that showed the most improvement, the senior management of the plant had fully embraced the new operating philosophy and developed an operating system that was suitable for the plant's culture. This led the former Toyota managers to conclude that simply understanding the system is not enough. If the culture of the organization is not capable of changing, then implementing aspects of the TPS might make the company more efficient, but it cannot fix the fundamental problems within the organization.

My goal for writing this book is to help the reader understand the basic concepts of the TPS and then how to put together a systematic process in the organization that will drive overall value for the organization. Through my many years of implementing the TPS in a range of operating environments, I have developed a methodology that incorporates the foundational principles of the TPS and incorporates them in a way where real value can be created from day one. I have no desire to create a group of companies that are working to clone themselves after Toyota.

The fact that many industrial giants today are eagerly pursuing and applying aspects and philosophies of the TPS would lead to immediate advancements in productivity over conventional mass production methods of yesteryear if it were not for the fact that most industrial giants are already doing a host of things correctly. The truth is that companies that have survived into the twenty-first century are already doing many things the best way. The concepts that I deliver in this book, although based on the TPS, have been enhanced to drive short-term, immediate, bottom-line impact in any organization.

2.4 Kentucky Alchemy

One example of how Toyota implemented their system of manufacturing with success is Toyota's plant in Georgetown, Kentucky. Toyota entered a nonindustrial, rural area of the United States, where people had little or no experience in auto manufacturing, and built the largest and most profitable facility in Toyota's arsenal. Toyota completed a task that has inspired almost every major auto manufacturer to follow in their footprints. Prior to Toyota's facility in Georgetown, little was known concerning how the rural workforce would transition from the fields to the assembly line. What was discovered was that the workers in the South wanted to avoid the influence of the United Auto Workers as much as the manufacturers did. This combination of lower cost labor and work ethic that began on the farm was a combination for manufacturing excellence. Toyota used this formula to turn steel into gold and then used that gold to finance their worldwide expansion. This has not gone unnoticed by the other automakers that have come to be known as the "new domestics." Foreign companies now see value in manufacturing products in the United States for sale in the United States. This offers these companies the opportunity to take advantage of low-cost labor while at the same time addressing political and national sentiment for buying "American." Every major auto manufacturer has at least one plant in the South, and many have multiple facilities. At each of these facilities, especially the ones developed within the last ten years, the labor rate is pennies on the dollar compared with wages in the company's home country.

How did Toyota achieve success with Georgetown? Simply put, it was the combination of the concepts behind the TPS coupled with the relentless passion for success by the workforce in Georgetown. Although the TPS is a complex management system based on, and formed by, pure and simplistic ideals, TPS is *not* merely a system of building efficient automobiles; it is more importantly a system that builds efficient people. The system encourages individuals to develop creative solutions to everyday problems.

The managerial norm for an average American organization is a top-down management style in which executives, managers, and engineers have sole discretion in shaping the methods and vision of the organization (Figure 2.1). This system always has, and always will, produce enormous stress in the workplace; it creates an invisible but tangible dividing line between "management" and "workers." When the people responsible for carrying out the plan of an organization, or as I like to say, the people who "do the work," feel that they have nothing to gain if the company is successful, then the company is at a disadvantage. This is not to say the organization will not achieve some level of success, but they will never reach their maximum potential because they have failed to tap into their most valuable resource: their people. The other problem is that the direction for the organization is driven from the management and not from the

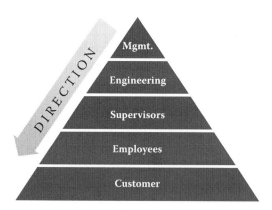

Figure 2.1 Conventional Management Philosophy.

customer. Management interprets what they believe to be the needs of the customer. This is great if the management team is correct; however, often the management team spends a lot of time adjusting the process to fine-tune what the expectation is for the customer. In a business like auto manufacturing, this can be costly since the capital necessary to retool the facility to make changes to the model can be costly. If management is wrong, then a popular product can lose the demand overnight and adjustments can take as a long as twenty-six weeks to implement.

Perhaps Company X has set a new goal to decrease the amount of labor hours needed to produce their world-famous widgets. Who better to determine where the savings should come from than the people who labor to produce it? For an engineer or a manager to go to the workers and determine how many labor-hours can be saved, and how it can be achieved, is nonsense. In these traditional top-down organizations, managers tend to stick to their desks and engineers to their tables; they are problem chasers instead of problem solvers. In these types of organizations, more often than not, the quickest solution is to just ask the workers to do more, or work harder. In Toyota there is a famous saying that embodies the philosophy of the TPS, and that is to "work smarter, not harder." Where would that leave Company X in terms of its long-range strategy? Does Company X even have a long-range strategy? How much harder do the workers need to work for the company to reach its goals? Unfortunately, this is an operating philosophy that is widespread across all types of organizations.

Contrary to traditional top-down management approaches, the TPS is based on a management system where the customer drives the direction for the organization (Figure 2.2). In Toyota this is known as "customer first." Note that the direction inside the organization flows from the team members, or the ones doing the work, down to the engineering and management of the organization. In this model the role of everyone in the organization is to support the people doing the work in order to provide the customer with the level of product that they demand. When you think of it, the closest person to the customer is the last person to touch the product. Generally this is a team member on the floor. No matter how well a part is designed or how smart the CEO is, if the operation does not execute at all levels then the customer will never gain the intended benefit.

When considering where the value is added in the organization, the only people who actually provide any value to the customer are the people making the product. Everyone else in the organization provides no value to the customer. It is only through this type of thought process that the

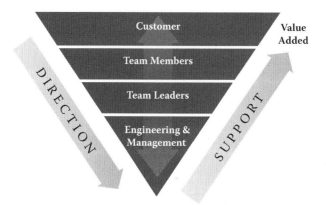

Figure 2.2   Customer First Management Philosophy.

maximum value can be achieved in the organization. Minimizing the people in the organization who do not create value allows for the organization to minimize costs.

2.5 Keep It Simple

To understand the TPS is to understand basic principles of simplistic management. Toyota has built a reputation as a leader in the auto industry for building quality automobiles at the lowest possible cost, but that is not the factor that drives the organization, it is just the desired outcome. The focus

Figure 2.3   Work Smarter, Not Harder.

of the Toyota organization, as a company as well as a culture, is on the workers, known inside the Toyota Management Systems as *team members*. Toyota has made itself successful through *kaizen*, or the small, continuous improvements in every aspect of the production process. The best ideas, or kaizens, do not come from members of management, but from the team members themselves. One of the strengths of Toyota is that Toyota believes in and encourages the ability of its team members to solve even the most complex problems in the organization. Toyota discovered in the early 1950s that the most valuable resource rested inside the minds of the employees, and it began developing a continuous improvement process in order to tap that resource of the organization and turn it into an innovation engine. That process was the TPS.

The first thing you will learn inside an organization that embraces this philosophy is that the leaders of the organization recognize that all members, from the president to the line workers, are team members first. Each member has the same responsibility to the organization. I always like to say it like this: "We are all members of the organization first; we all just have different roles to play within the organization."

Another way to look at it is like a team of rowers (Figure 2.4). Each person in the boat has a defined position. Based on position, each person also has a designated responsibility. As long as all rowers work together and fulfill their responsibility in the boat, the boat is able to be maneuvered successfully and ultimately will reach its destination. If one person in the boat decides to do his or her own thing, the boat becomes less stable and the progress of the boat is restricted.

When the boat is in calm waters and the rowers are not in sync, the progress of the boat is impeded. Although the boat is not operating at peak efficiency, the lack of synchronization by one member is more a nuisance than a danger to the other rowers. The boat still is able to avoid

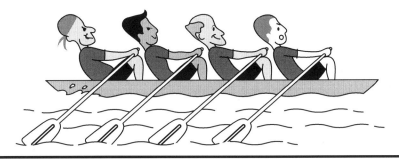

Figure 2.4   Rowing in the Same Direction.

danger along the journey since the calm waters provide an environment where the rowers have enough time to compensate for the rower who is out of sync. As the speed of the water increases and the boat begins to navigate more treacherous water, the response time to the dangers in the river is reduced. Anyone that has been whitewater rafting knows that you have to be aware of the dangers that lie ahead, and the rowers have to begin positioning the boat even before they approach a rapid if they hope to navigate the river successfully. If the rowers are not in sync when they begin to navigate the faster water, what was merely an impediment in the calm water can spell doom for all of the rowers if they fail to navigate the boat effectively.

This is the same scenario in businesses today. When a business is not stressed, the margin of error is much larger than when a business is operating under stress. For example, if a plant is operating at 60% capacity and a particular process is operating at 50% efficiency, the operation has to work 100% of the available hours to produce the required volume. Although this is a problem, the impact to the business is increased labor dollars for running the extra hours. Since the business is not operating at capacity, the plant management notices the variance in labor on the P&L (profit and loss statement) but don't foresee a major problem. When the business is not stressed, the natural tendency for management is to think that they have time to solve the problem. This generally is not a problem as long as it is identified and the management team has a plan to address the issue; however, more often than not, the inefficiency becomes the norm and the problem continues.

The next month the customer increases orders by 30%. The senior management in the company are ecstatic and have already started modeling how the increased sales will fall to the bottom line of the P&L. As the plant begins to ramp up production, they realize that they can't produce the required product because the inefficient process is not able to produce the increased components. Even though the customer is demanding more products, the business is unable to capitalize on the increase in orders due to one process being out of sync with the rest of the operation. Unfortunately this is a situation that I see every day as I work with stressed operations.

In an organization that is truly in tune with achieving success, all team members must work together to achieve the goals of the company. If the company is successful, then the employees will have long-term job security and this will help them to be successful. Regardless of whether it is Company X or Toyota, all companies share the same chain metaphor: they are only as strong as their weakest link. If Company X wants to improve the cost of the widgets they produce, they have to take an analytical approach to understanding where the problem exists. In the case of Company X, they are looking for ways to lower costs of the widgets, so naturally they need to look at all of the costs in the organization.

Organizations are complex by nature. The word *organization* insinuates that there are several things that when they act alone hold no real value but when organized they create an organization that has value. In this way businesses mimic nature, and the simpler the processes, methods, and procedures can be kept, the better the organization will be able to adapt to change. This concept is consistently demonstrated in nature where the simpler the life form is, the more adaptable it will be to the environment. Many relatively small organizations create such a complex structure that they feel burdened by the system to the point of inactivity.

For example, I was once working with a small distribution company with annual revenues of about one hundred twenty million dollars. We were in a meeting with the CEO, CFO, vice president of operations, and head of marketing. The purpose of our meeting was to try to identify why there was such a large percentage of our product inventory that seemed to not sell well. I had conducted an inventory of the top ten items that had the lowest sales in the warehouse, and we were going through each item to discuss how we could reduce the inventory.

The first sign of a problem came when the physical count of the inventory that I had made did not match up to the financial report. I discussed with the management team whether there was an accounting mistake or whether I had miscounted the product. To verify the count, I suggested that the five of us go out to the floor and count these ten items for ourselves so we would know exactly what the situation was. The CEO said that he had a better idea (I always love to hear the CEO say this): we should just have IT rerun the report and see if the discrepancy was an error in the system. The CEO said that it would take about six hours to run the report. I told the CEO that we are only ten feet from the floor and it would take us less than ten minutes to confirm this for ourselves! Needless to say, we went to the shop floor and checked the inventory ourselves.

This situation is a classic example of how some people overcomplicate even simple tasks. I call this *organizational paralysis* (Figure 2.5). The CEO had trained himself and his organization to believe that the organization was more complex than it really was, and therefore a simple process is made complex. Later in the same meeting, we were discussing why there

Figure 2.5 Organizational Paralysis.

was one particular item that we had only sold one of in the last twelve months. It sold for forty-five dollars and only cost us ten dollars; however, to get the item for ten dollars, we had agreed to buy one thousand units. When I suggested to the head of marketing that we should eliminate this product from our lineup, all she could say was that this was a great item, it had huge margins! I could not believe my ears. I explained to her that we were carrying ten thousand dollars worth of inventory to make thirty-five dollars in profit! She said, "When you look at it that way, it doesn't make sense, but we got such a good deal."

Even though these may seem like outrageous examples, they occur every day in organizations around the world. When I look at how to control spending in a company, I like to look at it like a person who is on a fixed budget going to the grocery store. If you have a fixed budget of one hundred dollars to spend, the first thing that you do is make a list of what your needs are. Generally when I make my list, I separate the "need" items from the "want" items. When I am at the grocery I pick up all the "need" items first, and then if I have any money left over I will purchase items from the "want" list. Typically it makes sense to make a list of what we need and only buy from that list. If I get to the grocery store and they have a special on corn, I don't buy a hundred dollars worth of corn when I only need one can. In this situation the discount of the corn has no value to me because there are other items I need and I can't buy them because all I have is corn!

Another scenario that I frequently run into is when I ask a question and the management gives me the answer that "it's complicated." This tells me that the only thing that is really wrong is the process within the company. If the organization has the control of the process, then the organization should strive to make the system as simple as possible. This allows for abnormalities to be readily observed and understood. Unfortunately this is not the case in many organizations. Many relatively small businesses (less than one billion dollars in revenue) overcomplicate themselves. Numerous times I have had spirited discussions with the CEO of a business concerning how the company, through internal systems alone, is overcomplicating its own situation. Sometimes we need to step back, look at the overall situation, and determine where the organization stands. *More often than not, if a manager sees something as confusing, then team members, and often the customer, will be confused as well.* Making things as simple as possible for the people on the floor, or the people creating the value in the organization, and providing them with the support they need is the true purpose of management in successful organizations.

2.6 The Toyota Production System versus Lean Manufacturing

Often I am asked, "What is the difference between lean manufacturing and the Toyota Production System?" Many people teaching and consulting on lean manufacturing today have a basic misunderstanding of the TPS, which, in the end, can only have negative effects on the organization. It is like having an incorrect recipe for baking a pie. There is a well-known story of how Loretta Lynn met her husband. She had entered a baking contest and the pies were sold to the highest bidder. Although Loretta had substituted salt for sugar by mistake, her soon-to-be husband bought the pie anyway.

If we are to understand how to drive value in our organization, we need to have the correct recipe. Although some may be able to swallow a piece of pie in which salt has been used instead of sugar, only love will let them eat it and smile. In business we can't make decisions based upon emotions and therefore we need to make sure that not only does it look like a pie but it also has to taste like a pie. To answer the question what is the difference between lean manufacturing and the TPS, the answer is in how you define each of them. There are many people claiming to understand the TPS who have no real understanding of the system and therefore do not get the desired results. I have met enough "senseis" in the world of manufacturing that I can see how business leaders can find the experience confusing and ultimately frustrating. Many people working in the world of lean manufacturing consulting have never really worked for a company where the system was implemented with any degree of success.

When selecting a lean consultant, or hiring a lean professional, it is important to understand where that person's experience comes from. Although there are many good organizations with a foundational understanding of lean, selecting a lean practitioner can be a challenging task. Just as there are people who are teaching lean manufacturing with a complete understanding of the TPS and can help your organization become very successful, there are at least as many people working in the industry with no real capability to help your organization at all. Even if the consultant has years of experience within Toyota, this does not mean that the person has a deep knowledge of the TPS. As with most organizations, not everyone in the organization has an equal understanding of the basic principles and the ability to transfer his or her knowledge into your organization. Therefore, choose wisely when selecting any consultant but especially a consultant that specializes in lean manufacturing.

One time I was restructuring the business of a seventy-million-dollar contract manufacturer. Generally I spend a lot of time working with the senior management in the organization to develop and implement the restructuring plan. The CEO researched my background and seeing that I had spent eighteen years with Toyota he decided that he would hire a plant manager with lean manufacturing experience. I did not have the opportunity to meet the person before he was hired, but the CEO assured me that he was a true "lean guy."

When the new plant manager started, I gave him a few weeks in the job before I scheduled a visit to his plant. Since I knew what the plant looked like before he arrived, I would be able to judge his knowledge of lean based upon what he worked on first. When I arrived at the plant the first thing he did was lead me into a conference room—mistake number one. As we sat in the conference room my knees hit something under the table and when I looked down I was surprised to see a calculator on the floor. The calculator had Velcro on one side and had been attached to the underside of the table. Puzzled, I asked if there was a purpose for attaching the calculators to the bottom of the table? The new plant manager spoke up and was beaming with pride as he explained that they often had meetings in the conference room and he had noticed that everyone always needed calculators so he bought calculators and attached them above every seat. I thought to myself that if this was the type of problem he had been solving he was spending too much time in the conference room—mistake number two.

As we continued our introductory discussion, the plant manager said that he had prepared a PowerPoint presentation that he would like to share with me. I thought okay; maybe he is going to show me some of the things that he had been working on in the plant; this day may be salvageable yet. As he started his presentation, he explained that this presentation was his philosophy of lean manufacturing. He had all of the basics of any lean story, but he made some strange additions here and there. For example, when he discussed 5S, he explained that he had added a sixth S, Safety—mistake number three.

I am always leery when people make up their own lean principles. I once had a long discussion with a colleague; he believed there to be eight types of waste rather than the seven types recognized with the TPS. His eighth waste was the waste of human ingenuity. When we had the discussion, I thought to myself that the eighth waste was actually the time that I was spending discussing his eighth waste!

Just when I thought I could take no more of the plant manager's presentation, he showed me a picture of an opossum that a road worker had painted over in the road. The next picture he showed me was a picture of him in the road picking up a dead opossum! Although I understood what he was trying to say, I was able to conclude right then and there that this guy was more than a "lean guy." Needless to say he wasn't the plant manager for long.

If lean manufacturing stays pure to its roots that are founded in the TPS, then the two are interchangeable. Unfortunately, in the world of lean manufacturing, that is generally not the case. The difference between lean and TPS is that in lean the focus is on the tools and with TPS the focus is on the system. There are many tools that can be utilized to implement the TPS, but they are not mandatory. A good example of this is seen in the example I shared of the plant manager who wanted to add the sixth S. Who can argue with the fact that safety is so important it should be the sixth S? The problem is that he didn't understand that implementing TPS is about the system. Safety has its place in the system and that is within the confines of standardized work, which is one of the foundational elements. It is these small distinctions that separate the real understanding of the TPS with the students of lean manufacturing.

My goal is to equip the reader of this book with the ability to understand the basic foundational elements of what is known as the TPS, and more importantly, some techniques for implementing the system with success in his or her organization.

The majority of misconceptions concerning the TPS originate from the sheer number of tools Toyota has developed for its implementation. Many organizations implement tools such as kanban and poka-yoke and realize substantial benefits in a very short time. Unless they have a broader understanding of the system, the one thing they will never realize is the true potential of the organization that comes only from a comprehensive understanding of the TPS. Selecting specific elements to implement is not necessarily bad; it is just limiting and will ultimately lead to frustration.

2.7 Standardization

I spent over eighty-eight weeks at Toyota being trained on various aspects of the TPS and the Toyota Way. Part of my training regimen at Toyota City, Japan, was to learn the basic foundational principles of the TPS. In Toyota, these foundational principles are represented with what is referred to as the Toyota Production System House (Figure 2.6).

The foundation of the TPS house is standardization: standardized work, jigs, tools, equipment, and locations for those items. Without standardization, there can be no *kaizen*, or continuous improvement; without standardization, the house of TPS would collapse. If there is one area where I see the most organizational opportunity, it is in the area of standardization. Upon the foundation of standardization rests the two pillars that support the house, *just in time* and *built-in quality*. Kaizen, or continuous improvement, is the roof of the house.

Through the years I have seen many different types of TPS houses. Many of the houses are complex, interwoven with many different threads. At the core of all of the houses used to represent the TPS are the same foundational elements; however, many people want to add to the house. Although I think it is better to maintain the basic simple structure of the house, as long as the understanding is correct, I have no real preference for which house people want to refer to. As we just discussed concerning organizational complexity, the rule of "simpler is better" should apply.

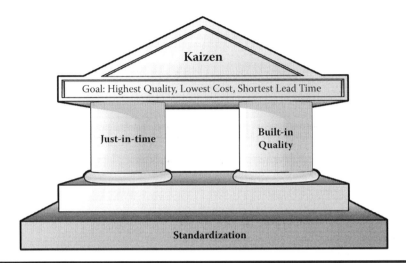

Figure 2.6   The Toyota Production System (TPS) House.

Referring to the house in Figure 2.6, standardization is the foundation for the complete production system. In Toyota, it is unthinkable to establish a process without first establishing standardized work. Standardization is the base of any good operational company. Many times I meet plant managers who try to convince me that their situation is more complex and different than anything that I have seen before and therefore it is impossible for them to have any form of standardization. I always find this interesting and insightful into the minds of the leaders of the organization.

Standardization applies to products, processes, systems, and procedures. Prior to any improvement opportunity, standardization must be achieved. Without standardization in place in an organization, it is like building a house upon the sand. As each day passes, the sand shifts and changes and can destroy any improvements that have been made. In any building construction project, we want the foundation of the building to be strong and immovable. Therefore, selecting the correct position for the foundation and making sure that the foundation is developed correctly are essential elements to any construction project.

When we think about standardization, the one tool that comes to mind above all others is standardized work. Many people misunderstand the concept of standardized work as only having application for manufacturing businesses. Standardized work is fundamentally a method of achieving repeatability in any given process. Whether it is a manufacturing process or a service-oriented process, the principles of standardization are the same. Every business is developed utilizing various business systems and practices.

If these systems and practices change every time the process is utilized, then the organization is going to have a difficult time providing a consistent result from the process.

This concept can be seen in a process as basic as providing the monthly financial report. If the accounting and finance organization follow a different process each month, then the financial information will vary, fluctuating even when the data are available for senior management to review. If the financial results for the previous month are not available for fifteen to twenty days after the close of the month, then how quickly does this allow the management of the organization to respond? If the accuracy of the data varies from month to month, how good are the decisions going to be that are made based on these data? Every business is driven by the systems within the business. Standardization has relevance to every area of business.

When we look at standardization from the aspect of the operations of the company, standardization translates into real value for the organization. Whether your operation involves manufacturing a complex product or not manufacturing a product at all, the key to efficiency and to quality is to have a repeatable process. For example, if your process is a carefully controlled metallurgical process, like the process used in the aluminum die cast industry, there are many operational variables that have to be monitored and consistently applied to ensure the quality of the product. In this situation, standardization is not only beneficial but it is also essential for maintaining the repeatability of the process. In every manufacturing and operational process, operational parameters have to be maintained to supply the product to the customer. Developing standardized work is the key to controlling these parameters and ensuring the repeatability of the process.

Why do we want repeatability? Repeatability is essential in an operational environment. Repeatability of the process enables the process to produce consistent and reliable results. By developing a repeatable process, we provide the foundation for kaizen, or continuous improvement.

Figure 2.7, which illustrates the kaizen continuum, shows that prior to any improvement, or kaizen cycle, standardization is required. Standardized work enables the operation to clarify abnormal situations immediately. Standardized processes and procedures also enable the organization to maintain a consistent level of quality and safety in the process. In the continuous improvement cycle, the key to maintaining the improvement in each cycle is standardization. Without standardization, the TPS is literally without foundation and therefore could not exist. Many organizations do an excellent job of implementing continuous improvement initiatives; however, without a

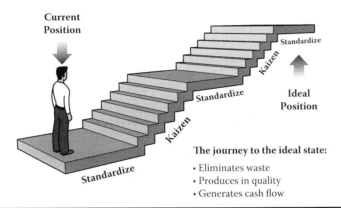

Current Position

Standardize

Kaizen

Standardize

Ideal Position

Kaizen

Kaizen

Standardize

The journey to the ideal state:

• Eliminates waste
• Produces in quality
• Generates cash flow

Figure 2.7   The Kaizen Continuum.

system that includes standardization, many of the gains are often diminished within six months after the initial implementation.

Maintaining standardized work is such a crucial element of the TPS that Toyota dedicates full-time resources to verifying standardized work at each process on a regular basis. In the same way that *shitsuke*, or discipline, is essential for maintaining workplace organization (5S), shitsuke is also essential for maintaining standardized processes and procedures. Many experienced practitioners of the TPS understand the frustration of implementing the five S's only to find the complete system falter due to a lack of focus on the fifth S. The same is true in the continuous improvement process. In many instances, the mere implementation of standardized work alone will yield tremendous operational improvements.

One of the most frequent problems with implementing any type of continuous improvement process is the lack of standardization prior to starting the improvement activity. No matter how simple or how complex the operation is, the first step to continuous improvement must be standardization. To facilitate the continuous improvement cycle, or the kaizen continuum as I refer to it, we must start with a standardized operation. Once the standardized process can be established, then we can start on the path of improving the operation.

By looking at Figure 2.7, we can understand this concept more completely. The current situation is represented at the bottom of the illustration and the ideal condition at the top. Now, to begin working toward the ideal situation, small incremental steps of improvement must be planned. After each stage of improvement, note the stabilization that is represented through standardization. This process continually repeats itself, with the next level of improvement only being attempted once a level of stability has been achieved from the previous improvement activity. It is this continual process that is constantly working toward the ideal state that is the kaizen continuum.

I have had the opportunity to work with many individuals and organizations seeking to understand and implement a production system based on the TPS. Unfortunately, many organizations focus their time and effort on the elements of the system and do not understand the basic foundational principle of the TPS: standardization.

Standardization is the foundation upon which the TPS relies to build a base that will yield sustainable results. Building a system without laying out the correct foundation can lead to some opportunity but will never lead to realizing the maximum attainable results. Implementing initiatives in a business requires energy from the management as well as the organization itself. This energy is infinite and therefore it is the responsibility of the leaders of the organization to make sure that the maximum return is generated from the energy expended to undertake any business initiative. To appreciate the importance of the role of standardization within the TPS, it is imperative to completely understand the principles of standardization. I define standardization as the method of producing a product, goods, or service by which quality can be controlled in the process repeatedly through thorough control of the variations in the process in order to produce the product or service within the desired specification and the time allotted.

No matter what type of industry or environment that we find ourselves involved with, we can always benefit from standardization. Without standardization, there would be chaos. If we think of our roads and highway systems, we can really appreciate the principle of visual management and standardization (Figure 2.8). I am continually amazed at the control brought to public transportation, just by virtue of a painted line.

Because of a few millimeters of paint on the pavement, vehicles can travel within a few feet of one another, creating order in a situation that could be

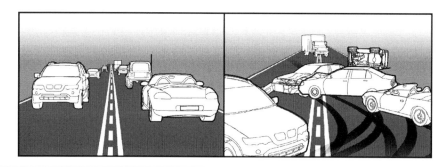

Figure 2.8   Real-Life Standardization (Highway Lines).

very chaotic. Although if you have ever driven in rush hour traffic in New York, you may question any concept of order; these lines are able to provide guidance for the operation of motor vehicles. If it were not for the painted lines on our highways, there would be mass confusion. The lines provide us with the basic information that we need to safely navigate the course ahead of us; there are lines that tell us when it is safe to pass another vehicle and when it is not. The same standardization and visual control lessons are just as valuable in organizations that are attempting to implement the TPS.

Without standardization, it is very difficult to manage the workplace or to even identify the abundance of opportunities for improvement. One of the main benefits of standardization is that it removes variation and exposes abnormalities. Just ask the six sigma black belts in your organization how much easier it is to identify and solve a problem when the task has previously been standardized. Without standardization the organization falls into chaos, with the quality of the product suffering as well as the overall efficiency of the operation.

Many companies resist efforts to standardize their methods and procedures because they feel as though it limits the creativity of the worker; this could not be further from the truth. Standardization levels the playing field and ensures that there is a basic understanding of what needs to be done and how it needs to be done. Although there may be many effective methods to standardize the processes and procedures of any organization, the premise that must be followed is that the standard is only as good as the organization's ability to follow that standard.

As stated before, standardization is the method by which quality can be controlled in the process. In manufacturing, the more a process is repeatable, the better the process will perform in regard to safety, quality, and productivity. Standardization also helps to control costs.

2.7.1 Quest for the Cube

I once conducted a kaizen event in a distribution center for a multimillion-dollar organization. The organization was going through a difficult time and we were looking for opportunities to reduce costs. When we studied the costs, we found that the transportation costs were running about 40% higher than the budget called for. The plant controller chalked up the increase to the rocketing cost of fuel prices.

When we began to dig into the problem, we saw that the controller was correct in reference to the total amount spent on fuel for the quarter (Figure 2.9).

Company XYZ Quarterly Report

2010 Quarter 1		
	Budget	*Actual*
Gross Sales $(000)	$57,486	$56,391
Return $(000)	$2,307	$2,209
Net Sales $(000)	$55,179	$54,182
COGS		
Material Cost $(000)	$44,371	$43,525
Labor Cost $(000)	$3,231	$3,169
Transportation Cost $(000)	$3,437	$3,675
Total COGS $(000)	$51,039	$50,369
Gross Margin $ (000)	$4,140	$3,813

Figure 2.9   Transportation Data.

Often I have found that people tend to think that most problems are out of their control. In this instance the controller was satisfied that he could explain the variance in the P&L by attributing the difference to an increase in fuel charges, something that was out of his control. Because fuel surcharges were a part of the transportation contract, the controller did not feel compelled to do any more investigation into the variance. Not satisfied with the answer from the controller and no doubt due to my training with Toyota, I could not help myself from asking why the fuel charges had risen so dramatically.

Company XYZ Transportation Detail

2010 Quarter 1		
	Budget	*Actual*
Total Transportation Detail $(000)	$3,437	$3,675
Fuel Surcharge $(000)	$1,467	$1,705
Fuel Rate $(000)	$3.02	$3.03
Fuel Volume $(000)	486	563

Figure 2.10   Transportation Cost Detail.

During our investigation, we discovered that the fuel rates had actually shown very little fluctuation during the two quarters in question (Figure 2.10). This surprised the controller and as we dug a little deeper, we found that the reason for the increase in fuel charges was directly related to the volume of fuel purchased. During the quarter, the quantity of fuel purchased had increased in direct proportion with the increase in fuel charges. By now everyone had caught on and was asking why this was the case when our actual sales for the period had shown a slight decline. What we discovered was very interesting. Even though our overall shipping dollars had reduced, the volume of delivery trucks had increased by 40%! Of course we were not done; we had only identified the cause for the increased fuel charges, and we still did not understand the reason for the increase in the volume of delivery trucks. In one respect the controller was correct in saying that fuel was a reason for the increased shipping costs; however, he had failed to notice that it was not the price of the fuel that was causing the increase; rather, it was the volume of fuel being purchased. As we continued to ask "why," we found that the reason for the increased number of trucks was due to a problem with the system that determined the cube of the truck.

Cubing, or utilizing the maximum capacity, is calculated for each truck and is controlled by a shipping program. Basically, the problem was that the trucks were shipping more air than products. To identify why we were sending out trucks without the proper load, we talked with the drivers and the fork truck operators, only to find out that both were following "standardized" procedures! We went to the shop floor to meet with the shipping supervisor and asked if he was aware that the number of trucks had increased and why.

Figure 2.11   Truck Cube Illustration.

The problem was that the standards were created specifically to show how the trucks should be loaded and unloaded; the actual cubic feet of the truck had not been taken into consideration when the standard was created. Since the overall volume of sales had remained constant, the problem was that the various sources of revenue had drastically changed, and the volume was made up of a different mix of customer orders. Although the trucks were running light (below capacity) on the customers with declining revenue, the trucks were blowing out (over maximum capacity) on the customers whose revenue had increased. This was causing the shipping and receiving supervisor to follow standard procedure for a blowout, and expedite the shipments to the customers. This was leading to even more costs as we were paying premium freight for those deliveries.

To fix the problem, we developed a mini cross-docking system that allowed us to make sure that the trucks were fully cubed before leaving the dock. To implement this system, we set up truck lanes in the shipping area and applied tape to the floor that had the exact same footprint of the trucks. Every truck was cubed out to the maximum capacity and was confirmed by the shipping supervisor. Some of the milk run deliveries had to be adjusted to account for the new customer mix. Although this problem cost us several thousand dollars, it actually saved us millions. Because we identified the cause of the problem, we were able to install a system that was flexible based upon the product mix we were producing. After that, we were able to standardize the system with a procedure. Now the process is adaptable and can meet the needs of the customers.

2.7.2 Visiting the Plant Floor

When I go into a company to make an assessment of where the opportunities are for removing costs, the primary thing I look for on the shop floor is standardization. Good operation managers will always take having a visitor as an opportunity to "spruce up" the plant floor so that it will show well. Standardization is something that no matter how much work the plant management does to prepare for a visit, I can see the real state just by observing the process. I do not know how many times I have had a plant manager read my bio on Monomoy's Web site, and in order to establish some common link he will tell me that he has a total understanding of the TPS. I even had one plant manager tell me that he went through Toyota's "TPS School" in Japan; it was not my place to tell him that a Toyota School does not exist. Although there are formal training programs, the real "Toyota University" is

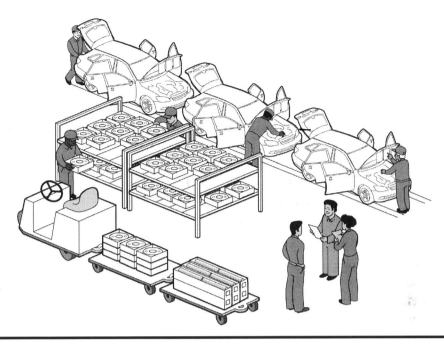

Figure 2.12   Supervisor Visiting the Shop Floor.

only found on the *gemba*, or shop floor. By understanding the plant's adherence to standardized work, I can determine how well the plant management really understands the concepts of lean manufacturing.

The worst thing that a plant manager can tell me is that he has a "complete understanding" of TPS, especially if the situation is one where we are looking to buy the company. I usually let these words go in one ear and out the other. The real indication of a plant manager's understanding of TPS can only be seen in one place: the shop floor.

When I go on a plant tour, I like being right next to the plant manager; I have developed a list of specific items that I am evaluating while visiting the factory floor. There are certain things that are observable and obvious to the trained eye; for other things, there are basic questions that I will ask the plant manager to gain an understanding of not only the process but the plant manager himself. One of the items that I am looking for is standardized work. I am not necessarily looking for standardized work charts; I am more observing the overall operation to determine if there is any level of standardized work for the operation. This consists of the overall value stream as well as specific processes. Whenever I ask the plant manager about standardized work, I usually get a response such as, "Yes, we have standardized work; everything that you see is standardized." Depending on the actual situation that I have observed, this can be a pretty good indication of whether the plant manager actually understands the basic elements of standardized work. More often than not, the plant manager is referring to the process by which the product is manufactured. Take casting as an example; the plant manager is fully aware that the casting process is standardized. Although this is true, when you look at the basic elements of the casting process, the standardized work exists in the details of the process. In this situation, I would rather hear the plant manager tell me that he or she has no clue about lean manufacturing, or TPS, than to claim he or she knows it all and is not practicing it.

2.7.3 Understanding Standardized Work

Standardized work is one of the most misunderstood components of the TPS. Since it is the very foundation of the success of the TPS, it is also one of the main reasons organizations fail when it comes to the implementation process. The error lies in the fact that small manufacturing companies are usually process-driven organizations; however, the process is not the production system, but the actual manufacturing process. For instance, in a casting process, the manufacturing process is fairly simple (see Figure 2.13).

Raw material is delivered in the form of billet, melted in a furnace, cast in a press or a die, trimmed, heat treated and inspected, and then packaged and shipped. This is a very basic manufacturing process and the organization places value on being able to follow this process from a technical standpoint to produce quality products. This process has operating parameters and standards for the metal composition, melt temperature, casting pressure, trimming standard, heat treat time and temperature, customer standard, and shipping quantity. Because this is an engineered process, there is a level of standardization that is engineered into the process. In this situation, when asking the plant manager about standardized work, it seems obvious that there is standardized work. The standards are clear and closely regulated to ensure that the final product is manufactured to these exacting standards. Although this is a necessity for the process to be successful, this is not what I am referring to as standardized work. I would refer to this as technical process control, not standardized work. Standardized work is the detailed process for

Figure 2.13   Process Flow Illustration: Casting.

manufacturing the individual products: how the casting operator loads and unloads the casting machine, how the material is placed after trimming, and so on. Many people who call themselves lean manufacturing experts believe that standardized work is only a document, the standardized work chart. This is far from the case. It is true that standardized work should have documentation, but the key to standardized work is the process, be it a person, machine, or robot.

Standardized work implemented correctly achieves repeatability of the process. If we look at the casting process again as an example, we can understand that even the best casting process is going to manufacture defective products; there are just too many variables in the process to be controlled; defects will inevitably occur. It is true that better organizations have better process controls, and they have fewer defects; however, they have defects nonetheless. The key from a manufacturing perspective is to control the abnormalities so that the customer does not see any fluctuation in the level of quality they are receiving. In Toyota, this is referred to as *jidoka*, or built-in quality. For this to be successful, each process in the production process has to achieve repeatability to ensure that the next process is capable of performing the required standardized work. This is followed from one process to the next with the final, predictable outcome of supplying perfect products for the customer.

Standardized work is the basis for creating the pattern of repeatability from one product to the next, one process to another. Through the utilization of standardized work, the entire production process can be managed. Without standardized work, waste will inevitably proliferate and flow from the top down throughout the organization. In a production process where standardized work is not present, waste readily shows itself.

2.8 Just In Time (JIT)

As we refer back to the Toyota Production System house in Figure 2.6, the pillars of the house are *just in time* and *built-in quality*. Just as the concept of standardized work is most often attributed to manufacturing processes, the pillars have application with all types of operations.

Just in time is the commonsense philosophy of controlling inventory by ordering and using the raw materials needed to produce only the products that are necessary to meet the order of the customer. When the concept is fully exploited, the desire is to operate with little or no in-process inventory, which results in shortened lead time as well as freeing up working capital. When properly implemented, just-in-time production ensures that what is being built is needed and what is needed is being built. The concept of built-in quality ensures that what is built is free of defects for the customer, and therefore all products that have been produced are converted to finished goods, minimizing work in process.

Consider the situation at 95% of all manufacturing companies today. All manufacturing companies have a process that takes a certain number of raw components and processes those components into a particular product that is then sold to the customer. The company only makes money when the product being produced can be sold to the customer. Any products that are produced in excess of what the customer is willing to purchase does not provide direct value to the organization today. All of the raw materials and partially manufactured components have no value to the customer, as they are only willing to pay for the finished products. Based on this situation, it only makes sense to minimize all excess inventories and strive to produce only what the customer is willing to purchase.

To achieve this process, the organization generally will prepare a production schedule that will drive the purchase of all of the raw materials and schedule all of the necessary production equipment and processes. The schedule is distributed to the related production departments within the organization and the necessary suppliers. The example would end here if the world were perfect and everything happened according to the plan. No changes, no adjustments are required in a perfect world. All of the suppliers are able to meet their commitments without a problem, the customer order does not have any adjustment positive or negative, and of course the manufacturing department runs as planned and everything is right with the world.

Even in sophisticated organizations such as Toyota, it is rare that the plan ever proceeds without changes and alterations. The reason for the change is simple. The conditions underlying the original production plan when it was first developed are destined to change prior to and during the time the plan is being implemented. Depending on the condition of the economy, these changes can often be quite severe; anyone who was in the automotive business from October 2008 through March 2009 knows this. Notwithstanding the radical changes that industry has seen during the great recession of 2009, it is an enormous task to coordinate the sometimes hundreds and even thousands of individual components that need to be manufactured to produce the volume of finished goods attributed to the production schedule.

Therefore, it is not surprising that often the production schedule has to be modified to compensate for the many changes that occur from the time the schedule is initially developed until the time the actual product is produced and shipped to the customer. The ability of an organization to manage these schedule modifications speaks to efficiency of the operation. When I am visiting companies for the first time, I spend a great deal of time observing the production environment. It is through these observations that I am able to identify how capable the organization is in regard to the overall operations of the company. When this process is not managed effectively, the symptoms are generally easily identifiable due to buildup of inventory, which increases the lead time for the customer and consumes working capital.

The concept of just in time avoids these problems by following three fundamental principles:

1. The Pull System—Buy only what is needed and produce only what you can sell.
2. Flow Production—Leveled production where the production is always moving.
3. Takt Time—Synchronized output based on customer demand.

2.8.1 The Pull System

The first fundamental principle of just-in-time production is the pull system (Figure 2.14). Through the years the pull system has gained a lot of notoriety in the manufacturing world. This can be attributed primarily to one of the tools utilized in the pull system, the kanban. Although the kanban is an excellent tool for implementing a pull system, it is simply a tool. Often I find in the vast ocean of materials on lean manufacturing and the TPS that there is a fascination with the tools utilized by Toyota and others for implementing just in time. Although this is not necessarily a bad thing, it is important to understand that these are simply tools and the system can be implemented utilizing various tools. Some people want you to believe that without the kanban, the pull system cannot be implemented. This is incorrect.

The real essence of the pull system is the flow of information. In traditional organizations, information is pushed through the system. Since material will ultimately follow the flow of information, the material ends up being pushed through the system, creating stockpiles of inventory at various stages of the manufacturing process. In Figure 2.15 the material is pushed through the system and there is thirty-six and a half days of inventory in the

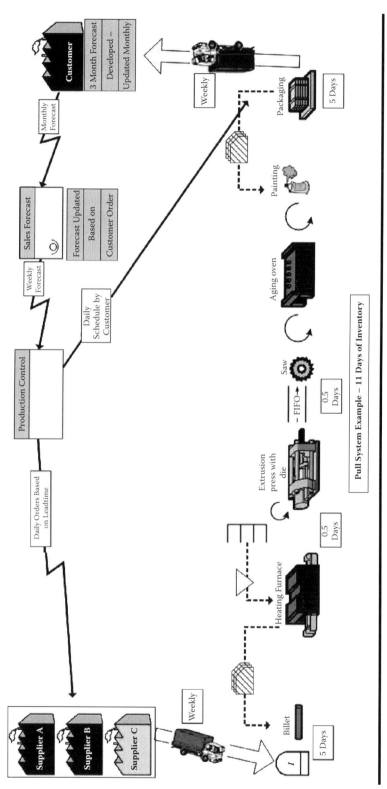

Figure 2.14 Pull System Example: Aluminum Processing.

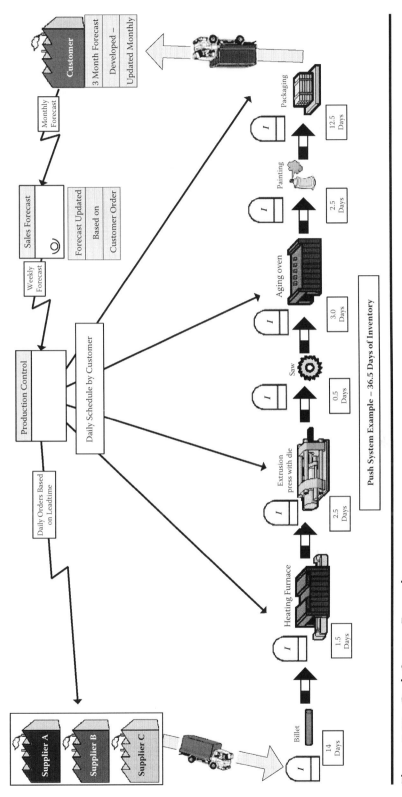

Figure 2.15   Push System Example.

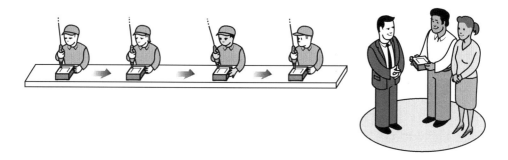

Figure 2.16 Pull System Example.

system. This is a lot of working capital tied up in product that has no real value to the customer.

In a pull system (Figure 2.16), the information flow is simplified. Without changing the requirements from the customer and for the suppliers, the overall flow of information can be made more direct and useful for the internal process. By using some simple just-in-time tools, including kanban, the overall manufacturing process is made more simplistic and inventory is controlled. The level of inventory is still not ideal, but it is much improved from the inventory level in the push system. The inventory level in the pull system is maintained at eleven days, which is twenty-five and a half days improvement from the push system. This not only simplifies the manufacturing process but it reduces the level of working capital needed to maintain the process.

Although this may sound a little complex if you are new to the lean manufacturing way of thinking, fundamentally under the push system you are producing whatever is processed from the preceeding process, and your ability to process the work in process from the preceeding process determines the level of inventory between the processes. With the just-in-time system, the preceding process pulls only the product necessary to complete the order. The process where the parts are pulled can then replace only what was pulled. By following this, some of the benefits of the pull system are the following:

- Excess inventory is eliminated.
- Production instruction is tied to the process that is closest to the customer.
- The production process is synchronized.
- Communication is improved between processes.
- The need for good quality and increased process reliability is highlighted, improving the overall efficiency of the operation.

2.8.2 Flow Production

The second fundamental principle of just in time is flow production. Continuous flow production is based on the concept of eliminating the stops and starts associated with manufacturing, thus keeping the production process leveled and maintaining the flow of the material through the process. Flow production works in unison with the pull system to reduce the overall manufacturing lead time and reduce the level of inventory in the process.

Ideally flow processing is achieved by producing product one at a time. To achieve one-piece flow, product is produced one at a time and passed to the next process. Producing parts in batches for the next process is not allowed under the concept of one-piece flow. By achieving one-piece flow, we can reduce the starts and stops associated with traditional batch production.

Let's consider Figure 2.17. In this example, it takes one minute to process each unit. Because there are four processes that need to be completed prior to the product being ready to be sold to the customer, there are four minutes of processing required to produce each product. Therefore, the production lead time for this product is four minutes.

Now let's look at a more traditional approach to manufacturing (Figure 2.18). In this example, product is still produced on an assembly line, but the products are produced in batches of twelve products. In this example, a unit still requires one minute to be processed through each of the manufacturing processes. In addition to the manufacturing time, each part now has to wait for the batch of twelve to be completed prior to moving to the next process. The parts wait an average of six and a half minutes prior to being processed to the next process. When we multiply this wait time by four processes, each product waits a total of twenty-six minutes and takes four minutes to process; therefore the total processing time is thirty minutes! This does not even count the waiting at the end of the process for the products to be palletized.

I am not saying that the TPS does not allow batch processing. The important factor to understand is that these are just concepts that are to be utilized as much as possible in order to get the process closer to the ideal state. Of course it is not always possible to produce the products one at a time. However, by utilizing the concept of one-piece flow, we can minimize the lot size of the batch, thus reducing the overall time that the part spends waiting for production.

The goal of flow production is to produce the products in the lowest possible lot that enables the product to be efficiently produced.

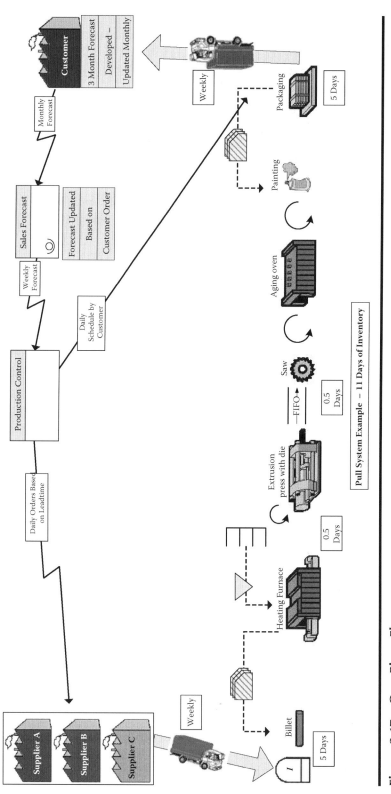

Figure 2.17 One-Piece Flow.

Figure 2.18   In-Process Inventory.

2.8.3 *Takt Time*

The final fundamental aspect of just-in-time production is takt time. Takt time is the synchronization of production based on the customer demand. Ironically, takt time has its roots in Germany. Takt comes from the German word *Takzeit*, meaning cycle time. This is interesting because cycle time and takt time are two completely different, though related, concepts.

Takt Time =	Total Daily Production Time
	Total daily customer requirement

Cycle Time =	Total Daily Production Time
	Total possible units produced

Takt time is the time that is necessary to produce one product through the production process. This time is taken by taking the total customer requirement per day and dividing this into the total daily production time available.

For example, let's assume that the customer's demand for a particular product is twenty thousand products per month. Since we have twenty days of scheduled production, this gives us a daily production requirement of one thousand products per day. If my production day is based on seven and half hours of production, that gives me four hundred and fifty minutes of production, which means that each product requires twenty-seven seconds to produce. My takt time for this product is twenty-seven seconds.

Understanding the takt time for each of the various processes is essential in order to determine the optimal process flow. If each process proceeds according to the specific takt time required, then only the number of products necessary will be produced. Takt time is the demand for the process.

Takt Time is the unit of measure for determing the demand for each product.

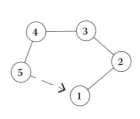

Cycle Time is the actual time necessary to complete a product.

Figure 2.19 Cycle Time versus Takt Time.

Cycle time, on the other hand, is based on the capability of the process. Each process must be able to complete the required cycle time to meet the demand, takt time. Often the cycle time is not controlled, and this results in overproduction.

A good example that illustrates the importance of takt time versus cycle time occurred when I was working on a project at a bakery. The project was to optimize the assembly operation for chocolate cakes. Yes, this is a real example; and no, it wasn't assigned by my ten-year-old son.

Referring to the illustration in Figure 2.20, you can see that the chocolate cake line originally had five operators. The process started with the first operator loading the cardboard trays that the cakes sit on. It turns out that the cakes need help staying on the tray, so it is necessary to apply drops of corn syrup to each tray prior to assembling the cake. The operator loads the tray into an automated system for applying the corn syrup. Because only one cake sits on a tray for a three-layer cake, the tray-loading process was not very busy. In fact the operator spends the majority of his time waiting for the next operator to remove the tray and begin assembling the cake. This is a classic example of muda in the process, but we aren't discussing muda until the next chapter so I won't go on about that.

Another interesting aspect about this operation is that the corn syrup application used to be a manual process. The former management team had hired a "lean guru" to help them optimize their process, and one of the results of the improvement effort was to spend twenty thousand dollars

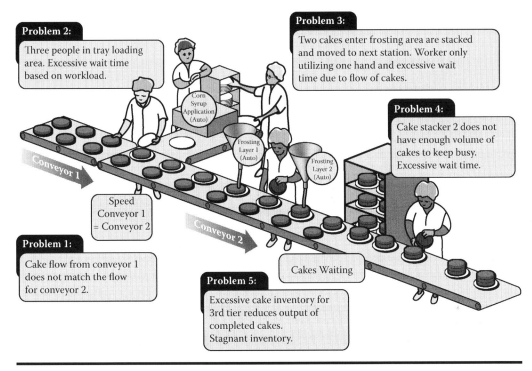

Figure 2.20 Chocolate Cake Line: Before Kaizen.

building a machine that would apply the two drops of corn syrup to the paper tray. I asked the floor manager how many people they had on the operation before, and he said they had the same number of people before and after the improvement process. I asked him what was the effect of the improvement, and he stated that the operator did not have to manually apply the corn syrup and this saved five seconds per process. If you can picture me scratching my head trying to figure this one out, it is actually quite funny. Again, this is a classic example of being wary of so-called gurus: not only was the process not improved, it was actually worse. The company had spent twenty thousand dollars to have the worker spend more time waiting. This is a classic example of good intentions gone bad.

The next process was responsible for putting the cakes on the trays. The cakes came out of the oven on a conveyor two at a time, and then the cake closest to the operator would be assembled to the cardboard tray with the corn syrup. At this process the cakes transferred from the oven conveyor to the assembly conveyor. The thing about this process was that the oven conveyor and the assembly conveyor were running at the same speeds. With the process being to assemble three-tiered chocolate cakes and the cakes running down the conveyor two at a time, this was bound to cause a problem somewhere along the line. However, when I spoke to the operator, everything seemed to be running smoothly.

The next process was the cake stackers. The frosting machines were on the same side of the line, and it was the job of the cake stackers to wait for the cake that was assembled to the cardboard tray to pass through the frosting machine and then assemble the other cake to the tray. This sounds simple enough, but this is where we saw the problem with the conveyor speeds. Since the cakes were fed two at a time from the oven and the assembly conveyor and the oven conveyor were running at the same speeds, where does cake stacker #2 process get the cakes to form the third layer of the cake?

The answer was the cake waiting area. It seems that every now and then when the cake stacker #2 needs cakes for the cake waiting area, the entire process is stopped and the cakes are removed from the assembly conveyor and are placed in the cake waiting area to be assembled as the top layer of the three-tiered cakes. During this process of abnormal handling, the cakes often would get damaged and there was not a good method for controlling the inventory or the process for restocking the cake waiting area. This led to old cakes being stored in the cake waiting area, and it also led to assembled cakes being stored in various locations. This complicated the process and made scheduling next to impossible for the next production line, which was responsible for packaging the cake.

Due to the fluctuation of the process, the supervisor had assigned a fill-in person to the area to help with the nonstandard work. This meant that we now had five people trying to assemble a three-layer cake with four processes.

Who would have ever thought that assembling a three-layer chocolate cake could be so complicated?

The good news is that even though there were a lot of complicating factors, the process was quite simple. Once the actual process for assembling the cake could be understood, we were able to improve the efficiency of the cake line by focusing on the takt time of the cake line (Figure 2.21).

The first thing we did was remove the twenty-thousand-dollar piece of equipment that the "lean guru" had installed. We reinstated the manual process for applying the corn syrup and combined the first two processes into one process. The second countermeasure we implemented was to understand where the production pace was controlled. Since the oven conveyor was fixed and the time to bake a cake was engineered based on the specified temperature and time the cake needed to be in the oven, we could

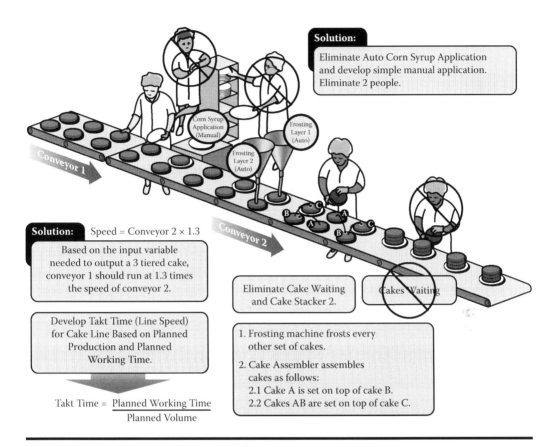

Solution:

Eliminate Auto Corn Syrup Application and develop simple manual application. Eliminate 2 people.

Conveyor 1

Corn Syrup Application (Manual)

Frosting Layer 1 (Auto)

Frosting Layer 2 (Auto)

Conveyor 2

Solution: Speed = Conveyor 2 × 1.3

Based on the input variable needed to output a 3 tiered cake, conveyor 1 should run at 1.3 times the speed of conveyor 2.

Develop Takt Time (Line Speed) for Cake Line Based on Planned Production and Planned Working Time.

$$\text{Takt Time} = \frac{\text{Planned Working Time}}{\text{Planned Volume}}$$

Eliminate Cake Waiting and Cake Stacker 2.

Cakes Waiting

1. Frosting machine frosts every other set of cakes.

2. Cake Assembler assembles cakes as follows:
   2.1 Cake A is set on top of cake B.
   2.2 Cakes AB are set on top of cake C.

Figure 2.21   Chocolate Cake Line: After Kaizen.

understand the process demand. We investigated to see if the cakes could run through the oven in rows of three versus rows of two, and this was not possible with the current equipment. The solution was to simply reduce the speed of the assembly conveyor. By reducing the speed of conveyor 2 so that conveyor 1 ran at 1.3 times the speed of conveyor 2 we were able to supply three cakes to the assembly conveyor and maintain an even supply of cakes. Since the line speeds had been adapted for a three-layer cake, the output was continuous and we were able to eliminate the cake waiting area. By reinstalling the frosting machines to a side-by-side configuration and establishing a pattern for the cake production, we were able to combine the two cake-stacking processes into one process. This virtually eliminated all of the waiting time in the process. Because we had leveled the production, we also had eliminated the nonstandard work and this eliminated the need for the fill-in person.

When you examine the new process, you'll see we were able to address all of the problems that existed in the process before the kaizen and while simplifying the process we were able to go from five people needed to assemble the cakes to two people. This was a savings of three people per shift.

2.9 Jidoka

Referring back to the TPS house in Figure 2.6, the other pillar of the TPS house is jidoka, or built-in quality. The philosophy of built-in quality is that quality is confirmed at each process, resulting in the finished products being defect free. Built-in quality can also be referred to as "customer first." Fundamentally the customer demands a product free of any defects; therefore it is everyone's responsibility to produce units free from defects. The only way to achieve zero defects, sometimes referred to as delta zero, is to ensure that each process has the ability to produce the level of quality demanded by the customer. From an operational perspective, the customer is always considered to be the next process.

To achieve built-in quality, the process has to have the ability to stop production whenever an abnormality in operation occurs. This is referred to as autonomation. Autonomation is different than automation in that the process, whether manual or automated, has the ability to identify the abnormality and halt production until the problem can be corrected. This can also be thought of as automation with a human touch.

There are significant benefits to building quality into the process. The first benefit is that defects stop flowing through the process, reducing the rework necessary as well as reducing scrap. This improves the efficiency of the operation. The second benefit is that the equipment is monitored more carefully, and abnormalities in the equipment cycle can be corrected prior to a catastrophic failure. This increases equipment uptime. The third benefit is that everyone becomes an inspector, and the need for dedicated personnel to just confirm the work completed by others is reduced. Finally and most obviously, this type of system exposes problems, thus making the problems easier to see. Forcing the problems to the surface allows management to focus on the issue and develop countermeasures, resulting in a more stable operation.

One example of built-in quality is the fixed position stop. Many factories have production conveyors, and a simple innovation such as the fixed position stop allows for abnormalities to be identified and corrected, minimizing the interruption to production. In Figure 2.22 of an assembly line, when a worker identifies a problem within the process, he pulls a rope known as an

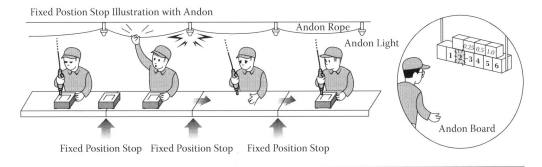

Figure 2.22   Assembly Line Andon Example.

andon rope. The andon rope activates a light known as an *andon light;* generally this is accompanied by an *andon tone.* As long as the process has not advanced to the fixed position stop, the andon light turns yellow and this activates a light on an *andon board.* The andon board is located in a central location, where it can be monitored by the supervisor. When the supervisor hears the tone, he looks to the andon board and can quickly ascertain where the problem is in this illustration, station 2. The supervisor can also see that the line has not stopped, since the andon board has a yellow light. The supervisor now responds to the point of the problem and works with the employee to resolve the concern. If the concern is resolved before the product advances to the fixed position stop, the supervisor will pull the andon rope to release the line, and production is never interrupted. If the product reaches the fixed position stop, the conveyor stops and the andon light and board display red lights indicating the line is down. This ingenious system is a simple and effective tool for building-in quality to the process.

Based on the foundation of standardization represented in Figure 2.6, and with Just in Time and Built-in Quality forming the supporting pillars, the roof of the TPS house is *kaizen,* or continuous improvement. The concept of continuous improvement is based on the philosophy of incremental improvement in the process.

2.10 Continuous Improvement (Kaizen)

Continuous improvement is not possible without a firm foundation of standardization in the organization. This is illustrated in Figure 2.7. Each incremental step in the continuous improvement process moves the process closer to the ideal state. Although each element of the house serves a purpose and they are all interrelated, the purpose of the TPS is continuous improvement. Without continuous improvement, the value of the system can never be realized. Continuous improvement is both a privilege and a commitment. If there is a "secret" element to the TPS, it is continuous improvement. Continuous improvement is a dynamic, ever-changing process. In Toyota we always used the saying, "You never measure yourself from where you were, only measure yourself from where you should be." Often management members tend to look at where they have come from and become satisfied; it is this process of always understanding the gap to the ideal situation that makes clear the opportunity for continuous improvement. Continuous improvement can also be very frustrating because it is like climbing a never-ending ladder. Senior managers need to balance the level of recognition for improvement with the desire for continually driving toward the ideal state. Only a healthy balance of each will motivate the organization to move forward. It was often very frustrating to work for Toyota because we continually measured ourselves from the ideal condition. This is why today you will find the senior management of every Toyota facility around the world discussing even the smallest margins of gaps to the ideal condition in their operations. Many companies would be completely content with an operational efficiency of 98.5%; however, it was this drive for the ideal situation that allowed the Toyota plant in Kentucky to achieve 100% operational efficiency in 1999, something that had never been achieved at any Toyota facility in the world.

2.11 Developing the Tools

What has become known as the great recession of 2009 had devastating impacts across almost every business segment. Many businesses that were on the brink of collapse prior to the recession collapsed completely, while the ones that have found themselves on the other side of the canyon are looking back wondering just how they made it. Currently there are many companies in a wide range of industries trying to adapt to the changed environment. Some of the companies are looking toward lean manufacturing and are searching for the tools that will enable them to improve their business. There are literally hundreds of books that have been written on the TPS, and there are many tools that can be used to make real improvement in any operation. The challenge that many businesses are facing is

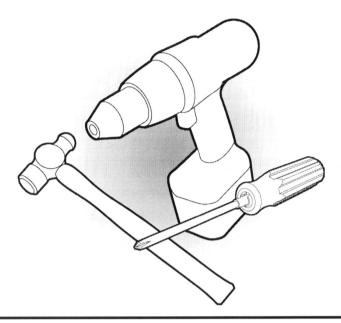

Figure 2.23   Tools.

that they do not understand which tool to deploy to make the improvement that they are searching for in their company. The many tools of the TPS are great tools; however, if applied without the knowledge and principles of the system, these tools can actually cause more harm than good. Many people within the realm of lean manufacturing are opposed to utilizing tools to derive short-term gains in the business. Personally I think that the right tools used in the right circumstances can be excellent catalysts for an organization. The organization fueled by the opportunity provided by short-term gains can often gain momentum toward longer term and sustainable improvement. The key is to understand the tools and to have the knowledge necessary to utilize the appropriate tool in order to capture the opportunity that is presented. Of course, for the organization to have long-term sus-tained improvement, the leaders in the organization need to have a more comprehensive view of how to develop a systematic process for continuous improvement.

Chapter 3

Why the Toyota Production System Makes Sense: Common Sense

3.1 Common Sense 101

In the fourteenth century, an English logician and Franciscan friar named William of Ockham introduced a principle that has become known as Ockham's razor. This principle states that all things being equal, the simplest solution is usually the best. This is especially true in manufacturing. The most common mistake I find in manufacturing, and especially operations, is that senior management wants to believe that their process of manufacturing even the simplest products is the most complex form of manufacturing and that if they have not thought of a solution then it probably does not exist.

A practical application of this point can be seen when we examine the now famous kanban system. So many times when I am talking to operating managers, they get caught up on the Japanese terminology and forget that the kanban is simply a tool for managing a very simple process found in manufacturing: the process of supply and demand.

When we think of the production process, fundamentally it can be broken down to the basic concept of the flow of material, or material flow. In every operating company, some type of material flows through the established processes, and some type of final product is produced and is consumed by the customer (Figure 3.1).

Often I am asked how I am able to apply tools that are fundamental to an industry such as the auto industry to a broad spectrum of organizations.

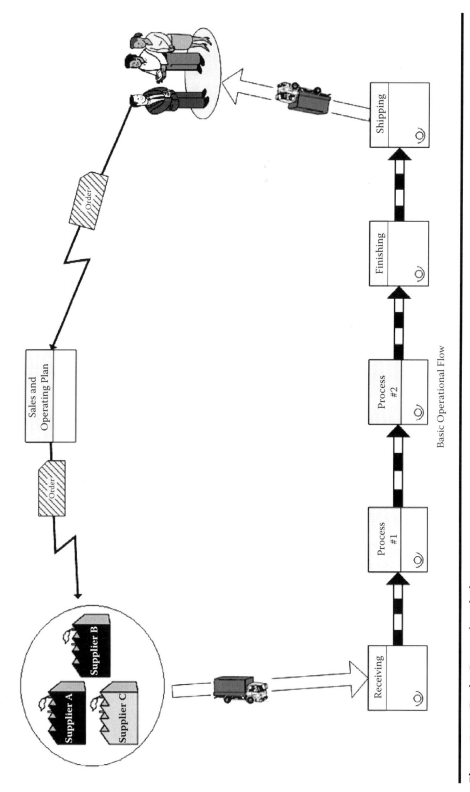

Figure 3.1 Basic Operational Flow.

The answer is simple: all companies operate in some capacity. It does not matter if the company is a manufacturing company or a service provider.

All companies receive something—let's call it a component—whether it is a raw material or information. Every company has a process and a plan to do something with, or to, the component to meet the needs of the consumer. In a manufacturing process, this is pretty straightforward because the components received are utilized to manufacture a final product based on the needs of the consumer. Every process in a manufacturing operation is developed and executed to manufacture the product necessary. In a distribution company, the component can be a final product such as a woman's shirt, and the process can be to receive that shirt from the manufacturer and then distribute it to the customer. It could also entail commingling that shirt with a pair of pants from another manufacturer and distributing the combined components to the customer. In both cases, the company is providing some type of service or value to the customer. The more valuable the process or service is to the customer, the more demand is created by the customer.

Why does the consumer choose to order the shirt from the distributor and not just order it directly from the manufacturer? Isn't it cheaper for the customer to buy the shirt directly from the manufacturer? The obvious answer is yes, but there has to be some value in the service that the distribution company is providing. In the example of a distribution company, the value is the economy of scale. Let's assume we are ordering a shirt from a manufacturer in India. If a shirt costs $5.00 to manufacture, the manufacturer has to make a profit so they build a 20% margin into the shirt and charge the consumer $6.00 for the shirt. Now I want to wear my shirt next week, so I need to ship my shirt by air. The shipping cost, once I arrange the shipping myself, is $3.95 to get the shirt to my house. Because I am importing the shirt directly from India, I also have to pay duty on the item and let's assume that represents $0.50. My total landed cost for the shirt is $10.45.

Cost of shirt	$6.00
Air freight	$3.95
Duty	$0.50
Total	$10.45

Now let's assume the value of the distribution company. The distribution company orders fifty thousand shirts and therefore the distribution company works out a deal with the manufacturer to lower the manufacturing margin to 10%, or $0.50. Now, because the distributor is shipping fifty thousand shirts, they ship them in advance of the season by boat, and because they are shipping so many other products, they are able to work a deal based on the scale of the shipment and the shirts are shipped for $1,000, or $0.10 per shirt. The distributor has to pay duty and because of the volume, the duty is much less, $0.25. The consumer orders the same shirt and the distributor has to cover the distribution costs and make a margin, so they charge 20% of the landed cost. The consumer has to pay local shipping of $1.50. Thus, the total landed cost for the consumer is $8.52.

Cost of shirt	$5.50
Freight	$0.10
Duty	$0.25
Dist Margin	$1.17
Shipping	$1.50
Total	$8.52

In this instance, the consumer saves $1.93. This is the value that the distribution company provides. This makes common sense.

From an operating perspective, if the process of distributing the product to the customer can be streamlined, eliminating waste, we can reduce the costs of distribution and we therefore increase the value the distributor provides for the consumer or we can maintain the price and increase the margin of the distribution company.

Although this is a simplified example, it illustrates how basic operations apply to all companies. Taking the time to understand the basic operation of any organization and then applying the basic commonsense principles of the Toyota Production System discussed in this book will lead to increased value in the organization. Whether the organization is faced with pressure from the consumer to lower prices, or whether the organization needs to reduce the overall operating costs to maintain competitiveness in the marketplace, these principles are a road map for improving the operations of any organization.

3.2 Understanding Value

Another commonsense principle that is reflected in the Toyota Production System that applies to every business is that the overall value of the business, product, or service is determined by the customer. Therefore, understanding the requirements of the customer enables the organization to focus on the correct things inside of the organization and maximize the value of the organization. The more valuable the organization is, the easier it is for the organization to remain viable. Especially during the challenges of a down economic cycle, these principles make more sense than ever.

Many organizations determine the price that the customer must pay for the product using the following method:

Material Cost + Operating Cost + Margin = Customer Price

This is the conventional way of looking at product cost. Using this model, the company sets the price, and the customer has a fixed price. Using this model is fine when the product is in high demand and there is a lack of competition. Using this model, the company can determine their margins, and therefore they have complete control over the cost of the product. What they do not have control over is whether the determined cost is in the range that the customer is willing to pay.

Another approach used in Toyota's system is for the company to have a good understanding of the cost that the customer is willing to pay for the product based on the value the product provides to the customer. In that case, we would follow this method:

Customer Price − Material Cost − Operating Cost = Profit Margin

As this formula illustrates, the customer, or the market, determines the price for the product; the profit margin is achieved by subtracting the costs. One of the benefits of this methodology is that because the market has determined the price, the margin can be increased by lowering the cost of producing the product. Because the company has direct control over the cost of the product, they can increase their margin by managing their costs. Although this makes common sense, this approach requires a lot of discipline for the company to be successful. Many times I have seen stressed businesses that are unable to

Figure 3.2   Customer Receiving Product.

contain their internal costs, and they are locked into a finished product cost with the customer. In these scenarios, companies get themselves into a negative margin situation. Often this is caused because the product requires a raw material with a lot of volatility in the price (e.g., plastic, copper, etc.). In these instances, it is imperative that the organization protect them from this fluctuation by putting into place pass-through agreements in the contract for raw materials or components that have a high degree of volatility. From a pricing standpoint, this makes common sense.

If we think of the basic fundamental of supply and demand, the manufacturing process needs to be able to produce the number of finished products that the customer is willing to purchase. When we think of manufacturing in this way, we naturally must start with the customer to understand how many products the customer is willing and able to purchase. This makes common sense.

3.3 Understanding Demand

Now that the organization has a good understanding of the price the customer is willing to pay, it is important to work with the customer base to have a good understanding of the overall demand. If there is one area of business

Figure 3.3 Excessive Products.

where businesses fail miserably, it is anticipated demand. Once the customer demand is understood, it makes common sense that the manufacturing process should produce only the products that the customer is willing and able to purchase. If the process produces more products than the customer is willing to purchase, then they will have purchased material and paid for the conversion of that material when there is no way to convert the finished products into cash. Therefore we can say that is common sense to produce only what is necessary to meet the customer demand.

To be able to produce the desired finished products, we are going to have to manufacture the required finished goods. To start this process, we need some method to tell the manufacturing process how many finished goods are necessary; this again is just common sense. Since we understand our manufacturing process, we also understand how material flows through our process and therefore we have to incorporate this information into our method of communication. For the sake of simplicity, let's consider that the easiest way to inform the production process what components are necessary is to write the information on an index card and give it to the delivery department. Let's call this index card a kanban. The literal translation of kanban is signboard (Figure 3.4).

Now that the kanban has been received from the customer, the delivery department can schedule the production based on the number of units that need to be delivered to the customer. The delivery department will take all of the manufacturing variables under consideration and determine when the products can be available to the customer. Because the customer is willing to purchase the products as soon as we can make them, it makes common sense that we should produce the products as fast as we can or with

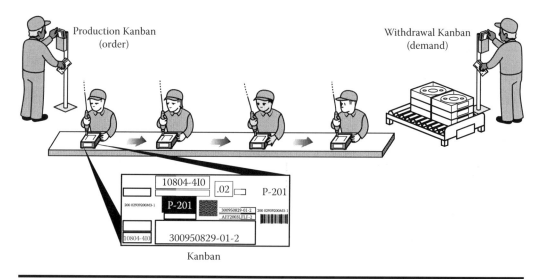

Figure 3.4 Simple Kanban Example.

the shortest leadtime possible. To do this, the delivery department will take the customer kanban and break it into production kanbans. The production kanbans will be based on the production capacity of the individual production processes and the conveyance time from one process to the next.

Now that the production kanban has been established, the pull process can begin. As discussed earlier, the pull process begins with the customer order. Once the customer order has been determined, then the delivery department will pull the necessary parts from the production process by issuing a production kanban. The production kanban will indicate to the manufacturing operation the number of products that are needed to be produced (Figure 3.5). The manufacturing operation will then issue supply kanbans to all of the raw material and component suppliers to begin producing the desired materials. Once the materials have been received, the production process will begin. Only ordering the materials that are necessary for production reduces costs for unnecessary materials. Although this process seems like a commonsense approach to manufacturing, many organizations struggle to understand how to manage this process.

Following this process ensures that the minimum resources are utilized, thus producing the product with the lowest available cost. This system of manufacturing has become synonymous with Toyota and is known as the just-in-time (JIT) method of manufacturing. As we discussed earlier, just in time is one of the pillars of the Toyota Production System House referred to in Figure 2.6. This is one of the driving principles that can be observed when examining Toyota's manufacturing methods.

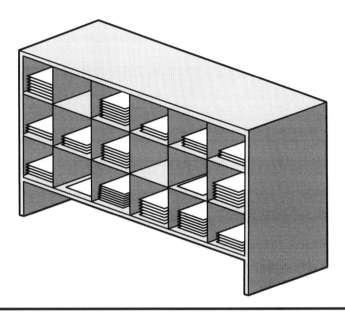

Figure 3.5   Heijunka Box.

Figure 3.6   Kanban in Use.

3.4 Common Sense in Action

Not only does this process make sense for a company like Toyota, I have seen firsthand how these principles apply to all types of organizations. It is only by taking a step back and observing the process from a distance that we can understand where the complexity of the process exists. As was illustrated earlier in the example of optimizing the chocolate cake line at the bakery, often the answers for improving a process are commonsense solutions to problems that others have spent a lot of time and money attempting to solve.

One example that is a vivid reminder of this principle comes from a CEO who attended a training program I designed in order to teach senior management and line management the basic principles of the Toyota Production System. Initially the CEO wanted to attend the training session to understand what all of this "TPS stuff" was all about. Entering the training as a skeptic, the CEO was surprised to see that during a two-week course of intensive training it was possible to make substantial improvement to the bottom line of a business by identifying and eliminating waste.

One example that was implemented involved the purchasing department and the fabrication department (Figure 3.7). While they were conducting a 5S exercise in the fabrication area, the CEO noticed that there was an unusual quantity of scrap tubing. Once they corralled and segregated the scrap, the

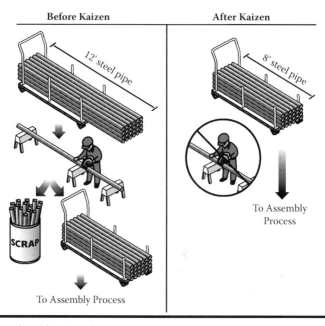

Figure 3.7   Excessive Pipe Sections.

majority of the scrap tubing consisted of four-foot sections of tubing. When the CEO asked the operator why there were so many sections of four-foot tubing being thrown away, the operator stated that the tubing comes to the facility in sections twelve feet long. The majority of the products produced require eight-foot sections of tubing, so four feet of tubing is removed from each twelve-foot section. The four-foot sections are too small to be utilized for other products, so they are sold for scrap.

Needless to say, the CEO was shocked that 30% of the material was being scrapped and sold at scrap metal prices. The CEO called a meeting at the process and brought the head of purchasing down to review the scrapped material. After some investigation, the purchasing manager found out that the company was buying the tubing at $2.50 per foot from the supplier. When the purchasing manager spoke to the clerk responsible for purchasing the material to understand why the tubing was coming in twelve-foot sections, they found out that they could get the tubing precut into eight-foot sections but that due to transportation, the tubing price would increase to $3.00 per foot. Because the purchasing clerk was asked to keep raw material prices down, he never considered buying the shorter tubing; after all, an eight-foot section at $3.00 per foot cost the company $24.00 while a twelve-foot section that cost $2.50 per foot was only $30.00. The company was receiving 50% more product for only 25% of the cost. What the purchasing clerk did not realize is that since the tubing was used in eight-foot sections, the company was paying $30.00 for a section that they could get for $24.00. Really they were paying 25% more for the material than necessary!

Although this seems like it is common sense, more often than not, this is exactly the type of situation that I experience working with companies of various sizes and degrees of sophistication. Often the most commonsense opportunity exists in the more sophisticated operations.

The CEO in this example was overwhelmed by the opportunity that was uncovered just by spending some time on the shop floor. Had the CEO not organized the 5S activity on the shop floor, it could have been months before the problem had been revealed. This is a perfect example of how the really valuable opportunities are literally hiding all around us on the shop floor.

Although these concepts seem basic and simple, that is the whole point of the Toyota Production System. The goal of any operation should be to make the process simple enough that anyone can come in and understand exactly what you are doing and why. More often than not, the management of the organization overcomplicates the operation to the point that they can't even tell you what is going on within the company.

Chapter 4

Common Misconceptions and Misunderstandings Regarding the Toyota Production System

4.1 TPS Misconceptions and Misunderstandings

I still find myself surprised at the abundance of misconceptions that surround the Toyota Production System. Of course, I started my eighteen-year career at Toyota as a production employee, so my education started from day one. By implementing continuous improvement initiatives throughout Toyota facilities and tier one suppliers, I have talked with a diverse cross-section of people, manufacturers, and non-manufacturers alike. Whenever we were working with suppliers, we found that they had a lot of perceptions pertaining to the TPS. Often people confuse the basic foundational principles of the TPS with the tools that are used to implement the system. They would come into the activity with the preconceived notion that TPS is a fixed system, that there is standardized work from start to finish on what to do, the equipment to do it with, and the manner in which it is to be done. Since just in time and built-in quality are really the foundational principles of the system, one *could* consider them as inflexible, but to what end?

Just in time and built-in quality are the driving principles behind *everything* that Toyota does. All of the tools previously mentioned are valid tools, but they exist solely to facilitate implementation of the system. If you strive to understand the core principles of TPS, it is inevitable that you will gain a comprehensive understanding of the outlying principles as well. If the tools are used without the core principles behind them, TPS ceases to be a system and becomes a short-term operational exercise. This is not necessarily a bad thing; it just should not be considered implementation of TPS or lean manufacturing.

The real questions we must consider are the following: *What is the best way* to ensure just-in-time delivery? *What is the best method* to build quality into the process? There is no correct solution for every business. Whatever your product is, the goal of your company should be to identify the best way to manufacture the product or to complete the operation. By asking ourselves what the best way is to manufacture the product, we can begin to understand what is necessary to achieve the ideal condition. It is only by measuring ourselves to the ideal condition that we can understand what opportunity exists to improve the organization. The key to implementing the TPS is to understand and manage the expectations of the organization. Understanding that there is not a "silver bullet" that will instantly transform the organization and that change comes through incremental small improvements helps the management team to frame the implementation and set the appropriate level of expectations.

At Toyota, for example, changes occur as a result of *thousands* of small kaizens implemented by the employees in their area of responsibility. Toyota values this system so much that they have developed a global system to capture these "suggestions." Employees are rewarded for each suggestion. Through this system, Toyota is ensuring that there is a systematic approach to capture the ideas of the workers. As each increment of improvement is implemented, the process moves closer to the ideal situation.

People with a misunderstanding of the true essence of the system often find themselves focused on the tools and not on finding the best way. These people look at TPS as a fixed system, with specific rules that must be followed (Figure 4.1). They see many of the tools as mandatory and the system as fixed and inflexible. This is the problem with many of the people who call themselves experts. This is even a problem inside of Toyota, with the numerous members of the management team who do not have a deep knowledge of the production system.

I have a much different way of illustrating the true essence of the TPS. I call this the *fried egg* illustration (Figure 4.2). The core, represented as the yolk, is fixed. The yolk contains the pillars of the TPS, just in time and built-in quality. As long as these philosophies are respected, then the system is flexible as long as you are working to determine what the best way is. The system is flexible to allow for the utilization of some tools, while other tools

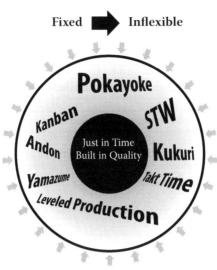

Figure 4.1 TPS Misconception: Fixed Tools.

may be ignored altogether. The key to understanding the *fried egg* illustration is to understand that some tools may make perfect sense to implement exactly how a TPS textbook might tell you to, and others may have to be altered based on the environment of your organization. By having the correct mind-set, we can rest assured that selecting the right tools is more important that trying to use every tool. Actually there are so many tools

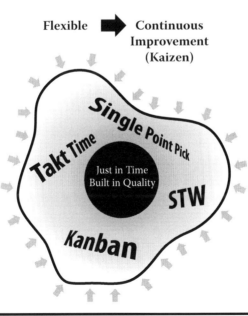

Figure 4.2 TPS Fried Egg Analogy: Flexible Tools.

used, it is impossible to conceive a scenario where this would be practical. A wise man once said that having skills is important, but knowing when to use them is even more important.

4.2 Which Way Is Best?

During the time I spent at Toyota Motor Corporation in Japan, I had the opportunity to be trained by a wide variety of company masters who lived and breathed the TPS. On one of my many trips to Japan, I had the privilege to undergo an intensive training session at the hands of one of the company's rising experts in all things TPS. The training session was made up of many long sessions at the facility, where we would eat breakfast, lunch, and dinner and work sixteen-plus hours a day. It was during one of the late-night sessions that I found myself sitting at a table taking a break with one of the senior masters in Toyota's training system. I remember it vividly because I had grown up in Toyota looking to this person for guidance. I had just been promoted, and we sat at the table with equal rank in the company although separated by the great gulf, which was his many years of experience.

I was at a point in my career where I had enjoyed some recent success in the company and some notoriety for developing a manufacturing methodology known as line simplification. Even though I knew that the master had much more knowledge and understanding, I thought that I had earned a sliver of respect from him based on my diligence to understand all things TPS.

As we sat at the table drinking our drinks, he looked up at me and asked me a simple question. He said "John-san, which statement is correct: (1) TPS is the best way, or (2) the best way is TPS." At the time I was eager to impress the master and spoke without thinking deeply. Instinctively I responded that TPS is the best way. There I was, sitting at the table with a

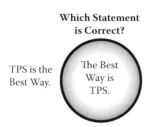

Figure 4.3   Question from a Master.

true master of the TPS, on the floor of the world-renown Tsutsumi assembly plant, working for none other than Toyota, whose production system was the envy of every other auto manufacturer; of course I thought TPS was the best way.

Needless to say, I was wrong!

The best way, the Master explained, is TPS because the essence of the TPS is the pursuit of identifying the best way of doing something. There are no predetermined solutions for every given situation; simply put, finding the best way to do whatever it is that you are trying to do is the essence underlying the TPS.

4.3 Toyota Production System Historical Overview

Toyota started in a small Japanese town named Koromo, in south central Japan, in 1937. By the time the town's name was changed to Toyota City in 1959, they had seven manufacturing facilities, all within thirty miles of each other. After the war, Japan was looking for some company to lift them out of the severe economic crisis that was plaguing the country. Toyota and the concept for producing automobiles seemed to be the right company at the right time. Even though much of the population could not afford to purchase an automobile, many outsiders were traveling to Japan to help with the rebuilding of the war-ravaged country. Supported by the United States, the entire infrastructure had to be rebuilt. Coupled with the fact that prior to the end of the war, all production was focused on supporting the war effort for the imperialist Japanese government and the majority of plants had been destroyed, the country needed some method for providing logistical transportation to the foreigners aiding with the rebuilding effort. This was solved with the introduction of the taxicab in Japan. Initial production of vehicles produced taxis for use by foreigners in the rebuilding efforts or trucks for use by the occupying American forces. Because of the lack of logistical capacity in the country, Toyota built up the supply base to produce their popular Crown taxi cab at the Motomachi assembly plant in what is now Toyota City, Japan. All of Toyota's suppliers were within thirty minutes of the plant.

In the meantime, Kiichiro Toyoda had arranged a visit to Ford Motor Company in the United States through the occupying American forces. The commanders felt it was better to help increase the domestic production capacity to produce the vehicles necessary for the rebuilding effort rather than expand Ford's manufacturing operations in Japan. Accompanying Kiichiro on this visit was a young and respected engineer named Taiichi Ohno. Taiichi worked at the Motomachi assembly plant and had responsibility for the engine machining operations.

During the visit to the United States, Taiichi marveled at the assembly line being used by Ford to produce so many vehicles in such a short time. The only point lacking seemed to be the level of quality being produced, as the Ford models seemed to be plagued with various manufacturing and engineering defects.

While on the tour of the United States, the team of Japanese engineers visited a supermarket (Figure 4.4). The engineers marveled at the many different products available to consumers from various food processors. The most remarkable part of the system was how the American supermarkets never seemed to run out of any one product. This was a huge problem in pre- and postwar Japan. The secret to this system of replenishment was a pull system that was determined based on customer demand (Figure 4.5). If customers were purchasing more corn than green beans, the corn was replenished more often. The system was controlled by reconciling the inventory at the point of purchase, thus triggering the reorder for that particular product.

Taiichi Ohno and the Toyota engineers made several successive trips to American supermarkets, even renting a home in Los Angeles to study this methodology more deeply. Out of this study, Taiichi Ohno developed what would later become one of the pillars of the TPS, just in time (JIT).

Figure 4.4   Supermarket.

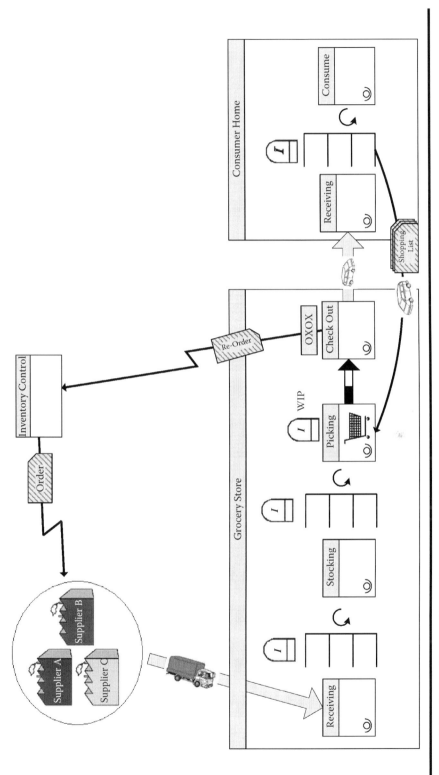

Figure 4.5   Value Stream Map: Grocery Store Process.

Since Toyota was under severe financial distress after the war, it was necessary to develop a system to increase production without expanding working capital. Even though the Toyota products being produced were in high demand, the banks would not loan money to a company that was dedicated to producing a product that the bank felt was not viable given the economic climate of postwar Japan. Given these financial constraints, Taiichi took the lessons learned from the American supermarkets and applied them to the supply base of Toyota (Figure 4.6). If the raw materials could be received by Toyota, converted into finished products, and sold to the customer prior to the due date of the invoice, then production could be expanded without receiving further credit from an unwilling financial market. This seemed to be the only method that would allow Toyota to expand their production capabilities. The first Just In Time (JIT) system at Toyota was set up in order for the sales cycle to be reconciled within the normal invoice cycle with the suppliers. At the time this was between 45 and 60 days. In order to achieve this a few things were necessary:

1. The products being produced had to have customers willing to buy them immediately.
2. The supply base had to be located close to the manufacturing plants as to not tie up inventory in transport causing excess levels of raw material inventory.
3. Production had to be continuous so that the raw materials being brought into the plant could be immediately converted into finished products.

To satisfy these requirements, Taiichi Ohno set about developing a tool for implementing JIT production. The system was based on the system discovered at the American supermarkets where components were pulled from their production locations based on need. Taiichi needed a method for signaling to the preceding process that the parts had been withdrawn and needed to be produced and, thus, the kanban was developed.

4.4 Kanban System Overview

Kanban, literally translated, means "signboard." Soon after the system had been developed, each production line at Toyota became a "supermarket" and supplied the next line (the "customer") with what they needed, when they needed it (Figure 4.7).

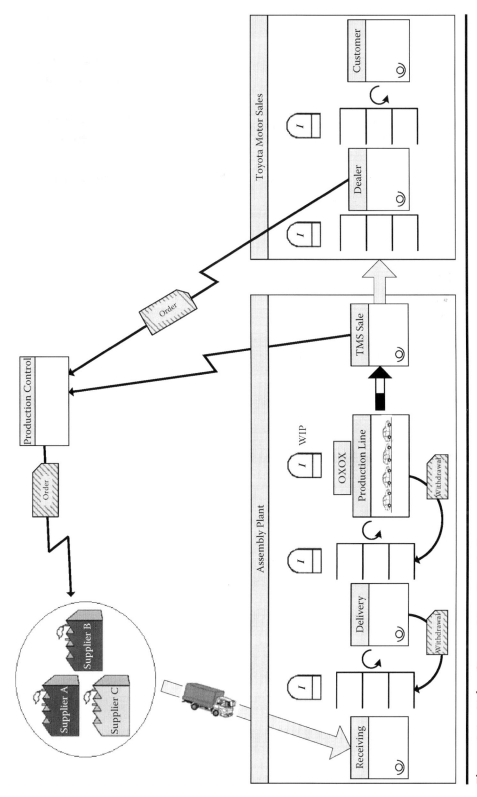

Figure 4.6 Value Stream Map: Toyota Process.

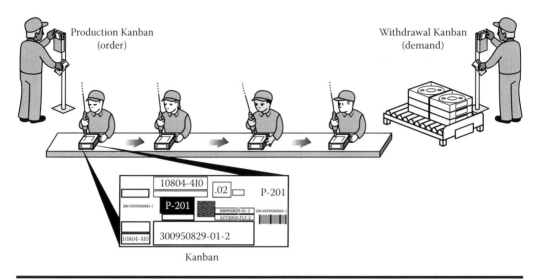

Figure 4.7 Simple Kanban Example.

In the kanban system, the first and most important factor is the demand. To satisfy the first requirement of the kanban system, only products that the customer was willing to purchase would be produced. To ensure this was the case, early Toyota salesmen would actually travel door to door, canvassing an assigned geographic area in order to understand exactly what the customers' requirements were. Once the demand level was known, it was necessary to have suppliers capable of producing the raw materials necessary to supply the Toyota factories. The postwar supply base was initially not capable of producing components in the desired levels, and this caused many situations where the line was stopped due to a lack of components. This violated the third principle of the kanban system of continuous production. The unreliability of the supply base was also affecting the quality of the products and because the finished products already had buyers, it was essential to have defect-free products in order to meet the demand.

Toyota quickly understood that for the kanban system to operate effectively, it would become necessary to educate the various raw material and component manufacturers in the principles of the kanban system. Suppliers that quickly adopted these principles were looked on as collaborative business partners, and businesses that could not meet the requirement were subsequently purchased by Toyota, and Toyota managers were dispatched to run these troubled companies. This was the beginning of what is now known as the *keiretsu*.

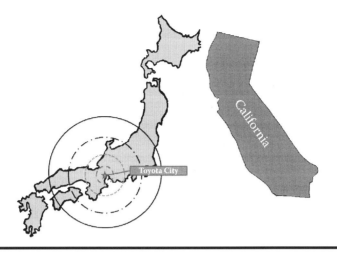

Figure 4.8 Toyota Supply Base in Japan.

Once the supply base had been stabilized, Toyota began to see the full benefits of the kanban system. Raw materials and components could be ordered, manufactured, and delivered readily to customers who had previously ordered units.

Understanding the history of the kanban system and how it was developed is essential to understanding JIT. Even though the kanban was an enabler of the implementation of JIT production, it is essential to understand that the kanban is just a tool for achieving JIT production and the kanban system itself is not JIT production. This can be better illustrated by understanding the implementation of the kanban system at Toyota's first wholly owned manufacturing plant in Georgetown, Kentucky.

When Toyota initially started production in the Georgetown assembly plant, many of the parts and components came directly from Japan (Figure 4.9). Often these parts were delivered in sea containers, and the entire purchasing and production scheduling was managed out of Toyota's Tsutsumi plant in Japan. This was necessary to get the plant up and running; however, the transportation cost to ship parts from Japan was high, and there was a lot of political pressure in the United States that in order for products, specifically automobiles, to be considered nonimport products, the majority of the parts and materials had to be sourced in North America. With the advent of NAFTA (North American Free Trade Agreement), parts produced in Canada and Mexico were given the same consideration as parts made in the United States. Once the supply base had been developed and the majority of components and raw materials were coming out of North America,

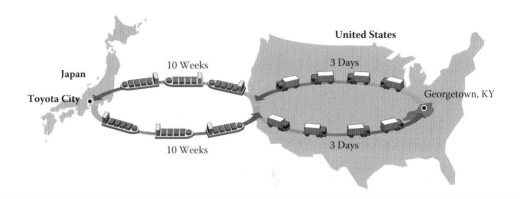

Figure 4.9 Toyota Japan Supply Chain to U.S. Operations.

the Georgetown plant began having problems with the implementation of the kanban system. Toyota sent over many experts from Japan to help determine the source of the problem and to get it corrected, as the unreliability of the supply was causing the assembly line to be halted frequently and was costing the company millions of dollars a day.

Initially we could not understand what was causing the problem with the supply. We knew that the problem was fluctuation, or *mura*, in the flow of the kanban; however, determining the exact nature of the problem was elusive. Everyone in Toyota was puzzled as to the nature of the problem because internally, our version of kanban was an exact replica of Toyota Motor Corporation's kanban system being used without problem in Japan. For several years Toyota would send people to the plant in Georgetown to "fix" the problem, and often people from the Georgetown plant would travel to Japan to learn the system, to no avail. Through the years, the problem was improved and the work stoppages decreased. Initially this was achieved by increasing the levels of inventory! This was considered taboo inside the company; however, with the lack of a solution for the fluctuation problem, the increased inventory was the only way to ensure that production was not interrupted. The cost of the increased levels of inventory was enormous, and even more pressure was applied from the headquarters in Japan to solve this problem. By this time, Toyota had started production at a new facility in Canada and that plant was experiencing the same problems as the Georgetown facility.

Finally, a team of American engineers using the most basic of tools in Toyota, known as the material and information flow map, or value stream analysis, identified the root cause of the problem.

It turned out that the source of the fluctuation was not the levels of inventory or the numerous other causes that had been identified (Figure 4.10). We

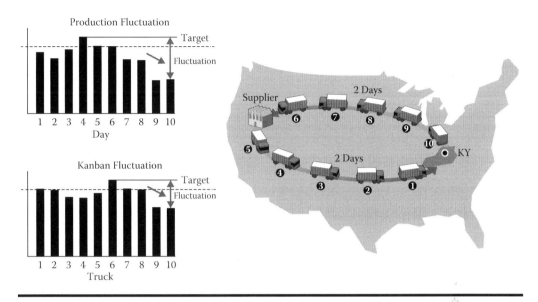

Figure 4.10   Fluctuation at Toyota; Georgetown, Kentucky.

had literally spent millions of dollars trying to solve this problem, and the problem ended up being one of simple geography. Given the fact that Japan is about the same size as California (Figure 4.8), the geography of the United States was wreaking havoc on Toyota's kanban system. The fluctuation that we were seeing in the kanban was created by the complexity of the supply base in North America. When Toyota originally started producing products in the United States, Toyota was a minority producer and had to utilize the existing supply base of the American automobile manufacturers. This meant that the supplier's location was dictated by either the proximity to the raw material or the proximity to the assembly plants of the American automobile manufacturers. In many instances, this meant that suppliers were hundreds or even thousands of miles from the production facilities (Figure 4.11). Any disruption that occurred in the production process was magnified by the distribution of kanbans in the system. If the plant had a low-volume day, then the kanbans returning to the customer would be less than the target production volume, causing a part shortage when that delivery returned to the plant.

Not understanding the full impact that the distance of the supplier to the assembly location was a critical error, considering that we had implemented an exact replica of the kanban system being used at Toyota's facilities in Japan. In Japan, the average Toyota supplier was less than thirty miles from the assembly plant. This allowed for quick response to any disruption in the operation.

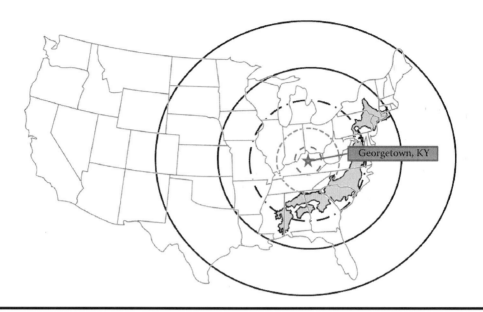

Figure 4.11   Toyota Supply Base in the United States.

Naturally when Toyota began operating facilities in the United States, it made sense to just copy the systems that had successfully been implemented and proven in Japan. When the kanban system was implemented, the system was set up exactly like a traditional kanban system in Japan.

We developed a system for North America that we called the e-kanban system (Figure 4.12). By electronically sending the kanban to the supplier, we not only eliminated the kanbans on the return truck, we also eliminated the number of trucks needed. The e-kanban system met with resistance and debate, even from within Toyota Motor Corporation, because of the elimination of the return truck kanbans; that was just not how it was taught and implemented in Japan. Once we implemented our e-kanban system, we saw remarkable improvement; by eliminating that fluctuation, we were able to reduce the overall fluctuation and the amount of materials needed; we even saved a substantial amount of money as well. Why? Because the best way is TPS and TPS is about searching for the best way of doing anything.

4.5 The Toyota Way

We have talked a little about the differences between TPS principles and the tools we have to carry out those principles. Many of the companies that try to implement TPS principles really just end up applying the tools.

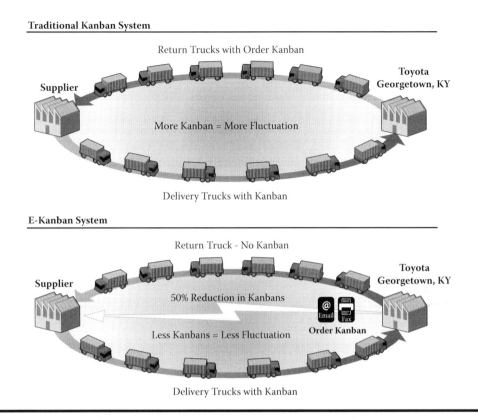

Figure 4.12   E-Kanban System at Toyota; Georgetown, Kentucky.

For instance, I could take the idea of kanban and implement it in a mass volume producer, make a few improvements, and save that company loads of money. I could implement the andon system or standardized work, and the quality would undoubtedly improve. In fact, this is exactly what I see in many of today's manufacturing industries, a watered-down version of the original.

The problem is, once improvement in an area has been quantified, the same great results are expected consistently with less time and effort, because their attention is already focused on another aspect of the plant. Eventually, there is an andon system in die casting, a kanban system in assembly, and standardized work in powder coating. Manufacturers will find themselves with all these TPS threads running through their operations with no clue as to how to tie them all together. I mentioned earlier how frustrating the Toyota environment can be at times; I can only imagine the frustrations faced by managers who implement TPS tools and have impressive, quantifiable production and efficiency results, yet still face the same bottom line problems. The headaches must be epic.

So, the question now becomes one of managing. How can a mass production company manage the tools created for a one-piece flow system? Is a management system needed to manage the system? Is it even possible for a system to manage another system, and if so, what lies at the heart of that system? How does Toyota do it? The answer is an easy one: manage it the Toyota Way.

The Toyota Way is not a system, process, or program; rather, it is a mindset wherein thought and action guide how we interact with one another and the way we manage on a daily basis. Managing the Toyota Way is centered on two principles:

■ Respect for people
■ Continuous improvement (kaizen)

Traditional organizations today, especially those in the manufacturing industry and corporate America, are structured with management at the top followed by engineering, supervisors, employees, and then finally, at the bottom, the customer. In this relationship, the employee is closest to the customer, yet the direction for products, goods, or services comes from management. The processes are then determined by the engineers, who then determine the work steps for the worker to build quality into the product.

The basic philosophy of these organizations operates on the assumption that the most important people are the ones at the top. Management makes all the decisions, and workers carry out the tasks. Often this philosophy will actually flow down to the customer, and management ends up dictating the product the customer receives. It is the modern equivalent of saying, "You can have any color you like, as long as it is black." In these organizations, there is no respect for the individuality of each person within the company. Although disrespect is not a conscious decision or policy for these companies, it manifests itself simply by their ignoring the priceless input that the workers can give. There is no method to capture ideas that will move the organization forward. Motivating the workforce becomes nearly impossible because motivation must come from the top and generally does not trickle down to the workers on the shop floor. A CEO's speech can get the front office really fired up and energized, only to be lost on the people who actually perform the work.

In traditional organizations, it makes no difference what direction management wants the company to take if the employees are not able to discern the fundamental purpose of what the company is trying to do. The greatest ideas for implementing any change in a work environment will fall flat unless the employees buy into it. Because of that exclusion, your customers will never realize any benefits of managerial organization, and your strategy loses all value.

4.6 The Customer Knows Best

From a business standpoint, we exist in order to serve our customers. Our customers are the only reason we exist. That is so important it deserves to be said again: *our customers are the only reason we exist.* Customers dictate what levels of quality and value they expect, and it is the business's responsibility to fulfill those expectations, period. Customers' needs must be put in front of the needs of the organization. By following that simple truth, we arrive at the conclusion that *the needs of the customers become the needs of the organization.*

Without customers, there is no business. That is one reason so many systems are in place at Toyota: they are there to understand *the needs of customers.* Without this basic understanding, it makes it difficult for any company to be customer focused, or to show that they hold respect for their people. Because every sequential process at Toyota is considered a customer, they are as important as the final customer. Because Toyota's line workers are the closest direct contact they have with their customers, they realize that without them, there would be no product for Toyota to sell. It is the workers who determine efficiency and quality levels by the simple fact that they show up for work and do their jobs on a daily basis.

4.7 Go. See. Act.

Genchi gembutsu (Figure 4.13) is the term used at Toyota when problems arise. It translates to "go, see, and take action," and its application is taken very seriously. Many of the people involved in the lean/TPS world confuse this term with *gemba,* which means simply to go and see. Although gemba is an important concept, its use is generalized for anyone in the company. Genchi gembutsu, on the other hand, is for those who solve problems by taking action, which is the heart and purpose of the Toyota Way.

As a leader in an organization, I might develop some great initiatives, but it will always be the employees who have to carry out the implementation; it will be the employees who standardize it. My role as a leader is to be in

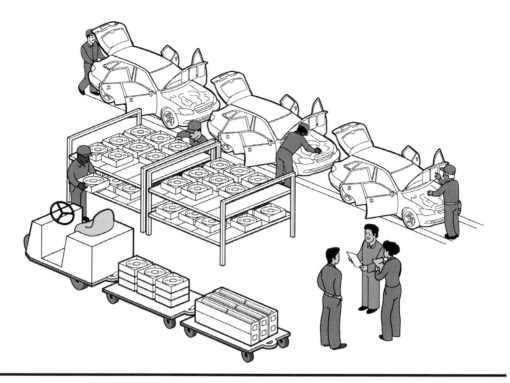

Figure 4.13   Genchi gembutsu (Go, See, Take Action).

touch with these people and to support them in our pursuit of the ideal condition. When the benefit of a great initiative becomes clear to the workers themselves, the benefit in value is soon seen by the customers.

Many companies today have genuine customer service policies; unfortunately, they only come into play once the product has been purchased by the end consumer. Because every sequential process inside Toyota involves customers, anytime they have a problem, it means the final customer has a problem. It is a leader's duty to make sure that that problem is eliminated. Taiichi Ohno summed it up best when he said, "No problem discovered when stopping the line should wait longer than tomorrow morning to be fixed."

Chapter 5

Waste Management ...
Improving the
Manufacturing Process
One Kaizen at a Time

5.1 Gap Management

The Toyota Production System was developed as a commonsense approach to improve productivity. It is those two words, *improve productivity*, that hold the key to so much about TPS and how it works. TPS is not specifically developed to build a better car, no matter how well built they are. TPS is designed with people, processes, and operations as the input, and quality as the output. The simplicity of the equation is the exact reason why TPS will work for anything. It is designed to produce common sense, not an automobile. Fortunately, Toyota does not have a patent on common sense.

The goal of TPS, in the broadest of terms, is to understand and implement the best way of manufacturing a product. The best way often can include automation as well as the human contribution to the process. The underlying genius of the TPS is that it is fluid and will work in any area of operations. One of the key objectives of TPS is to understand the current condition in relation to the ideal process. Once these items are clarified, we can determine the "gap." By understanding the gap, we can determine the path toward achieving the best way of manufacturing.

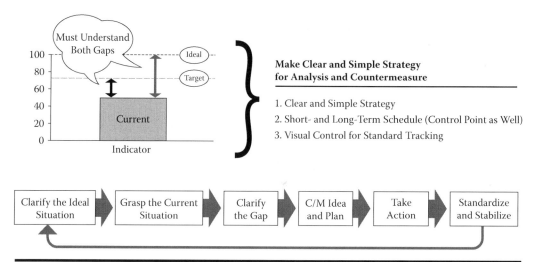

Figure 5.1   Gap Management Philosophy.

At Toyota, this philosophy is known as gap management (Figure 5.1). The first step of gap management is to understand the gap between the current situation and the ideal, or target, condition. This seems like a simple concept, yet only organizations and managers that have a firm understanding of the current condition will be able to use it. The purpose of gap management is to make clear and simple business strategies for analyzing the current and ideal, or target, conditions and developing countermeasures that close the gap.

There are six steps to implementing gap management (Figure 5.1).

Step 1: Clarify the ideal condition. Because the ideal condition is often unattainable, targets are routinely set that close the gap to the ideal state. Value stream mapping is an excellent tool for understanding the ideal state of an organization. Many people believe you should grasp the current situation before you clarify the ideal condition. This may be conventional, but it is not preferred. If we study the current condition prior to clarifying the ideal condition, we will constrain our thinking to the current situation. Average managers measure progress by measuring where they are from where they were. Great managers measure progress by measuring where they are and comparing that to where they should be (the ideal state).

Step 2: Grasp the current situation. Understanding the ideal state is something that many organizations and managers gloss over as a given. I cannot tell you how many times that I have sat down with heads of operations in a company or even CEOs, and they have no idea what is actually occurring on the shop floor. They often have a very good understanding of what they think should be occurring, but rarely is this actually occurring. Sadly, this is more the norm than the exception.

Step 3: Clarify the gap. Although targets may be established, since the gap is often vast between the current state and the ideal condition, both need to be quantified. Only by understanding both the ideal state and the target will you be able to understand the progress made once the target has been achieved.

Step 4: Develop countermeasure ideas and a plan to implement the countermeasures. When the gap is understood, a definable action plan to achieve the target must be developed. Often many countermeasures are necessary to close the gap. It is essential to understand the contribution of each countermeasure toward the target condition.

Step 5: Take action on the plan. No plan has ever solved any problem. People solve problems. No plan is self-executing, so it is essential that the plan is closely developed so that it can be executed. To execute the plan, clear responsibility and accountability for each countermeasure has to be clearly established when the plan is being developed. It is the manager's responsibility to make sure that the plan is executed.

Step 6: Once the ideal, or target, condition has been achieved, the process has to be stabilized and standardized. I have seen many great plans get executed and achieve phenomenal results, but they are not sustained and the results don't last. Another sign of a great manager is the emphasis he or she places on sustaining the results. It is only through sustainable results that real value can be created for the organization.

Once these steps have been successfully implemented, then the process is repeated. Because the process is never-ending, the organization continues to get better and better with each successive iteration of improvement activities. Even though the ideal state is rarely achieved, the gap management philosophy ensures that the organization is achieving the best possible condition for the process.

Most of the time the ideal way is not practical, and it takes people with knowledge of the operation to determine the best way. Toyota is filled with various experts on building cars, and yet many of them would have a difficult time entering your factory and telling you what will work best. The system of TPS respects the expertise of the people doing the work, and therefore it is essential for people with knowledge of the process to determine the best way.

5.2 The Three M's

For us to be equipped to implement TPS, one of the most foundational items is the ability to understand the areas of waste in manufacturing. In Toyota these are referred to as the three M's (Figure 5.2):

■ Muda—*waste*
■ Muri—*overburden/irrationality*
■ Mura—*variation*

At Toyota, we never told people to eliminate waste. Instead, we would encourage our people to *identify* waste. It seems only natural that once the waste has been identified, it will be eliminated. It is also true that waste cannot be eliminated until it has been identified. In this chapter, we will take a deep dive into understanding and identifying waste.

Figure 5.2 Three M's (Muda, Mura, and Muri).

5.2.1 Muda

In the most literal translation, muda is pure waste. However, muda can be organized into seven specific categories of waste that plague the manufacturing process. Determining the correct classification of muda is the first step toward developing a countermeasure that will reduce or eliminate the muda from the manufacturing process. The seven classifications of muda are transportation, waiting, overstock, overproduction, repair, overprocessing, and non-value-added work (NVAW). Even though waste by definition does not add value to the final product, some waste is necessary to complete the manufacturing process. In fact, it is only the proper understanding and classification of waste that allows us to minimize the negative effects and maximize the potential for process efficiency. After this brief introduction to the mudas, we will take a closer look at each one and discuss the best possible countermeasures.

5.2.1.1 Transportation

This waste is so obvious, many people do not consider it a waste but an essential aspect of business. Absolutely, it is an essential aspect of business; from paper clips to aircraft engines, everything gets moved around. Since transportation is necessary for moving products from one location to another, many manufacturing companies overlook this area as waste. From the perspective of the customer, transportation itself does not provide value.

When transportation is assumed to be essential, an opportunity for kaizen is lost. By classifying transportation as waste, we open up the opportunity to minimize the amount of transportation in the value stream of our product and process. Transportation can be one of the more costly forms of waste, especially when we consider the overall cost of delivering the product. From internal transportation (raw materials and subcomponents) to external (finished goods to customer), there are *always* opportunities to reduce transportation.

5.2.1.2 Waiting

Whether at work or in our personal lives, waiting too long for anything produces frustration. From the forty-five-minute wait at your favorite restaurant to the hours spent waiting at an airport, frustration can cause a host of problems. These frustrations manifest themselves in our personal lives in a variety of ways, from jaw-clenched finger tapping to the full-on irrational "snap" that finds us cursing the toaster. Whereas waiting in your personal life produces mainly intrapersonal frustrations and the occasional broken toaster, waiting in a work environment can not only produce external frustrations that affect you and the quality of your work but also has the potential effect of lowering morale and productivity. Waiting is one of the easiest forms of waste to identify. Almost anyone, when taken into a manufacturing environment and asked to identify waste in the operation, will point out people waiting as waste. One method I use for teaching people to understand the productivity levels of workers is to watch their hands and feet. It is really hard to work without moving your hands or feet.

5.2.1.3 Overstock

The first of our two "O's" is overstock. In terms of an operation on a production line, overstock is having more stock, or components, than are necessary to complete the process. Overstock hides problems in the value stream and costs the company additional operating capital. Overstock includes work in process (WIP) but does not include finished product inventory (FPI).

Of course, some level of overstock *is* necessary to account for fluctuation (*muri*), which we will talk about later.

5.2.1.4 Overproduction

This is where you find the just-in-case attitude as opposed to a just-in-time mind-set. Often referred to as overproduction, inventory is one of the hallmarks of traditional manufacturing processes and a common ailment of many American manufacturers. Overproduction creates many problems. Where do you store it? How do you control quality? When is enough enough? Who thought this was a good idea? Why do we do it? How much is this costing us? Many people confuse overstock and overproduction; here are a couple of ways to differentiate between the two:

Overstock is any work in process that is in excess of a production lot.
Overproduction is any *finished* product in excess of what is planned.

Another way to remember this is that all overproduction is overstock, but not all overstock is overproduction. Overstock will always be raw materials, subcomponents, and WIP; overproduction refers solely to finished goods.

Some inherent problems of overproduction include the possibility of hidden defects or contamination of finished goods that would require secondary processing. Another serious problem to consider is the fiscal stagnation from finished goods tied up in inventory for which the company is *not* getting paid. I cannot stress enough how important it is to clearly understand the differences between overstock and overproduction, because there are very specific countermeasures to be taken depending on the waste identified.

5.2.1.5 Repair

All repair processes are inherently muda. Repair is waste, pure and simple. Repair is also a good example of a necessary type of waste. In manufacturing it is inevitable that repair will be necessary. Even in the most efficient and quality-conscious facilities, it is not a realistic expectation that processes with multiple manufacturing variables, including human workers, can consistently produce vehicles without some level of abnormality. Some might ask why we even classify repair as waste if it is inevitable. If the waste (in this case, repair) is not identified, then creative solutions cannot be developed that can minimize this type of waste.

When we think about the seven wastes, Repair helps to think of it in terms of value; what is the customer willing to pay for? That question is what separates a value-added activity from a non-value-added activity. As we continue to discuss the seven wastes, we should keep that distinction between value-added and non-value-added wastes in the back of our minds. Keeping a focus on the simplicity of what *value-added* truly means can be of great help when identifying and classifying muda.

5.2.1.6 Overprocessing

The sixth classification of muda is overprocessing. Overprocessing is the work that is completed in excess of the work required to complete the value-added work (VAW) in a process. For example, perhaps our 60T hot chamber die cast machine consistently leaves flashing that must be removed before the process is complete. This is classic overprocessing. Overprocessing can also be more subtle in nature. Take, for example, applying a label; the work of removing the backing paper from the label would be considered overprocessing.

5.2.1.7 Non-Value-Added Work (NVAW)

The last of the seven types of wastes can be one of the trickiest to identify. The saying "you can't see the forest for the trees" is a good analogy to describe

NVAW. Many people get stuck in a rigid line of thinking that equates non-value as being non-essential. So, when a manager walks by and sees a worker sanding a cabinet with a random orbital sander, he looks at the overall process as essential, and therefore value-added. Although a smooth cabinet surface *is* essential, the manager completely overlooked the distance the worker traveled to pick up the sander, the time it takes to decide which sanding grit is appropriate, how the sanding disc is changed, and so forth.

Customers only want the finished product. They generally do not care one way or another about how it is produced. For example, we have a customer who pays for and wants a red car. The expectation of that specific customer is converted into dollars only at the moment the trigger on the paint sprayer is pulled; the expectation stops as soon as the trigger is released. That is VAW that the customer gladly pays for. What no customer gladly pays for is prepping the paint sprayer, donning protective gear, or any of the motions associated with the moments leading up to, or moments beyond, the actual act of painting. If we as manufacturers think of all those motions up to and beyond the paint process as VAW, then we miss tremendous opportunities for kaizen. Identify the work for exactly what it is; consider nothing as being too small or inconsequential.

These are the seven specific wastes classified as muda. If we want to eliminate waste, it is paramount that the waste be identified and classified properly; the countermeasures vary widely by each type of waste.

5.2.2 Muri

Our next *M* is muri, which is defined as overburden. Overburden occurs when workers exhibit more effort than required to complete the unit. Overburden could be as simple as a worker who continually has to deal with poor quality from the vendor, or it could come in the form of walking a longer distance than necessary.

Muri is often caused when management tells employees to just work harder. When a component part comes in to the wrong specification and it requires the worker to rework the part, this can cause the worker to bear more burden than intended. This is muri. Similar to the concept that the customer should only pay for the value-added portion of the process, the worker should only have to deal with the burden that is necessary to manufacture the product. This ensures that the level of physical exertion can be managed, and this enables the production process to be more consistent.

5.2.3 *Mura*

Our final *M* is mura, which is fluctuation, or unevenness, either in process or production. This is where Toyota really differs from other manufacturers. Understanding the nature of fluctuation in customer orders gives Toyota the ability to create stability within the internal manufacturing process. For instance, a customer places an order for the following cars: two red cars, one with power steering and air conditioning, the other with a CD changer; one green car with leather interior; three black cars, one with GPS, one with a sunroof, and all three with different engines. The TPS seeks to find the best way to level the production. In Toyota, leveled production is referred to as *heijunka*.

Although fluctuation generally manifests itself in the scheduling process, it is prevalent in the manufacturing process as well. Suppose we are producing a group of three products; one has a cycle time of forty seconds, one of fifty seconds, and the third takes sixty seconds to process. If the demand time, or takt time, for the finished products is fifty seconds, then our manufacturing and scheduling systems have to be able to balance the workload to ensure that the weighted average cycle time (WACT) is less than the required takt time. The brilliance of manufacturing is getting the WACT as close to the takt as possible without going over. This epitomizes the essence of what leveled production, *heijunka*, is working to achieve.

5.3 Classification of Muda

Now that the general introduction to the classification of wastes has been completed, a more detailed review of each of the seven types of muda, and specific countermeasures for each, is called for. For a comprehensive understanding of the seven wastes, see Figure 5.21, a muda summary chart, later in this chapter. This chart is an effective tool for identifying and classifying muda. It is only through the ability to correctly classify muda that a countermeasure can be achieved.

For general manufacturing organizations, the initial focus should be on understanding the three M's and establishing a manufacturing system that minimizes the impact on the operations. Although all waste is inherently bad for manufacturing, muri and mura can cause significant problems in the overall manufacturing process, whereas muda tends to manifest itself in all areas. Once the general understanding of the three M's exists across the

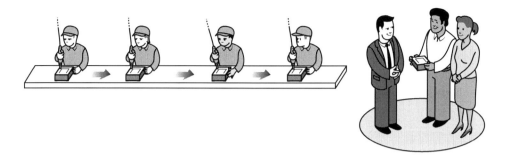

Figure 5.3   One-Piece Flow.

organization, it is beneficial to begin working to identify and focus on the seven types of muda. These are the most common areas for process kaizen.

Before we begin discussion of the seven types of waste, let's take a moment to examine the theoretical ideal manufacturing state, one-piece flow.

In Figure 5.3, the workers are each producing one product with no WIP in between processes and no stock of any kind. The customer is present at the end of the production line to take possession of the product, achieving what many would call the ideal one-piece flow. Although this may be ideal from a one-piece flow standpoint, it does not necessarily mean that it has the best manufacturing result. The goal of the Toyota Production System is not to achieve one-piece flow; it is to find the best way to manufacture the product. In some cases, one-piece flow is the best way and in others it is not; however, the concept is the driving philosophy behind identifying and countermeasuring waste. To optimize this process and find the path to the best way, we need to understand the three M's and the seven types of muda.

5.3.1 Transportation

Transportation, no matter how you move it, push it, shove it, or drag it, will always be waste. The only positive aspect to the waste of transportation is that it is so easy to see. When you see any part or pallet being moved with a fork truck in a factory, it is waste (Figure 5.4).

Transportation is necessary in a manufacturing environment; however, classifying it as waste will force the organization to minimize the transportation of the product. When I go into a plant and I want to understand the overall flow of the product, I often complete a material flow diagram, also called a value stream map. Many people, generally those who have not worked in a plant, advocate spending a great deal of time gathering all

Figure 5.4   Transportation.

of the details to complete the value stream map. In contrast, the necessary information to make a 90% accurate judgment can be gathered in a couple of hours of walking around on the shop floor. An example of this type of document can be seen in Figure 5.5.

As we examine the different classifications of muda, we will also see that they interrelate and often feed off of one another. For example, overstock and overproduction can cause transportation, and transportation can cause overstock and overproduction.

If we look to the transportation figure (Figure 5.6), we can see that this could be an example of many manufacturing plants. Finished goods are collected at the end of the production line, and once a full pallet of products

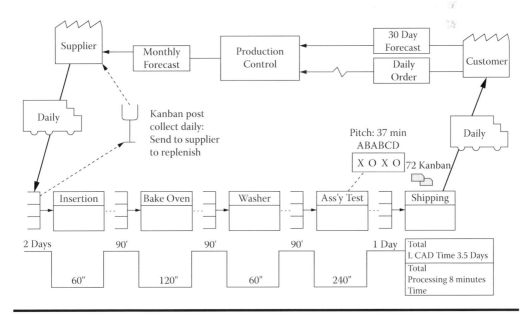

Figure 5.5   Hand-Drawn Value Stream Map.

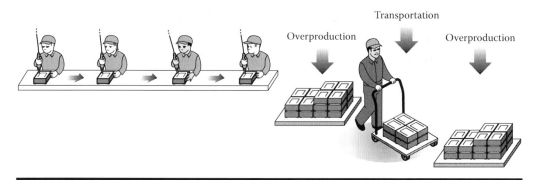

Figure 5.6   Transportation and Overproduction.

is completed, the parts must be transported from one area to another. This transportation is waste created by the overproduction of finished goods. Because the products have to be transported from one area of the plant to another, the transportation time now determines a level of overstock that is necessary to achieve a smooth flow throughout the operation.

As we look at the example shown in Figure 5.7, the quantity of product in the collection point has to be high enough that one pallet of parts is available every ten minutes in order to prevent the worker from waiting once the cycle has been completed. This also dictates that the warehouse has a minimum time of ten minutes plus one pallet to ensure that they can ship should the customer pull an order during the production cycle. To minimize overall waste in the operation, this has to be looked at holistically and not only from a transportation point of view. Some organizations prefer to minimize inventory and have many short cycles with small loads. This type of transportation system is referred to as high-frequency, small-lot production. Other organizations prefer to minimize the transportation and have fewer cycles with longer cycle times. This enables the worker to handle more products, but it also requires more inventory as the cycle times increase. This type of transportation system

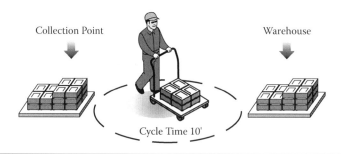

Figure 5.7   Collection Point.

is referred to as low-frequency large-lot production. There is no right or wrong answer; the best solution for each company depends on the priority of the organization, and what is best for each manufacturer. The one thing that should be constant is the ability to recognize and classify the waste.

When I worked for Toyota in Europe, we were faced with some unique challenges. The plant in the United Kingdom had gradually become isolated from the supply base. When the plant in the United Kingdom was constructed, GM and Ford had several manufacturing plants in the United Kingdom. As GM and Ford began to move their operations to mainland Europe, the supply base followed. This completely changed the logistics situation for Toyota. The cost for the transportation of parts across the English Channel had a serious impact on the operation. As the problem was studied, it was found that the cube efficiency of the transportation from mainland Europe was a source of concern. It was found that 30% to 40% of the shipments were not being utilized. In manufacturing lingo, this is referred to as "shipping air." As you can imagine, the cost for transportation from mainland Europe to the United Kingdom came at a premium. To decrease the volume of air being shipped across the water, a method for maximizing the cube efficiency had to be determined. As the situation was studied, it was determined that the implementation of a consolidation center in mainland Europe could enable the removal of fluctuation from container to container to be minimized. This operation would also enable the "milk run" logistics system to be consolidated for the broader European operations. (A "milk run" is a transportation system in which one truck will pick up parts from various suppliers before delivering the products to the consolidation center.) This consolidation center is commonly referred to as a cross dock. It may also be referred to as a warehouse. This may seem shocking to those who study TPS philosophy because many people believe that Toyota operates without warehouses to store components. However, it is exactly this type of process that is the essence of the TPS. The search for the best way often will lead an organization to consider unconventional methods.

5.3.2 Waiting

In Figure 5.8, an example of a simple production line, we can see that the third operator has no work and is idle. This idle time is referred to as waiting. In this situation, operator three is waiting on operator two to complete his operation. Waiting has no value because no VAW can be performed. Waiting also is an indicator of problems in the manufacturing process. If we assume that all of the operators have the same cycle time and the conveyor

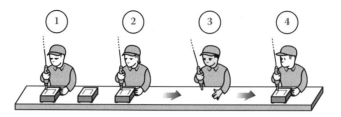

Figure 5.8   Waiting.

is controlling the takt time, then there had to be a problem with operation one or two in order for operator three to have no work. In a production environment where an andon system is in place, the operator with the problem would have stopped the line and all of the other operators would be waiting, until the line restarted. However, in a continuous line environment where an andon system does not exist, this would be a typical example of problems in the upstream process. The other problem could be that the work is not balanced and operator three completed his work early and is waiting for the completion of work from operator two. There is another type of waste shown in this scenario that lets us know that is not the case. It is the overstock between operator one and two. This indicates that the problem lies between operator one and two. By correctly identifying the waste, we can also begin the problem-solving process.

Waiting can manifest itself in several different ways. Wait time can exist for equipment as well as operators. It is always best if the operator and the equipment work in harmony and neither has idle time in the operation. However, as those who are experienced in manufacturing understand, the pace of the machinery doesn't always harmonize with the pace of the operator. As we search for the best way, which of the following situations shown in Figure 5.9 is better? In the first scenario, we have the machine with a longer cycle time than the operator. This situation is generally referred to as a machine-based cycle time. In the second scenario, the operator has a longer cycle time than the machine. This situation is generally referred to as an operator-based cycle time.

Which is best? The textbook answer is that neither scenario is ideal. Coming from an environment where I have never personally experienced the ideal situation, what do we need to know to determine what is the best way? To determine the best way, we have to make sure we have a thorough knowledge of the actual process. Understanding the scenario also enables us to identify the countermeasures to improve the overall efficiency and make the situation better. In this situation, it is important to understand the demand and then to calculate the takt time. If the machine cycle time can

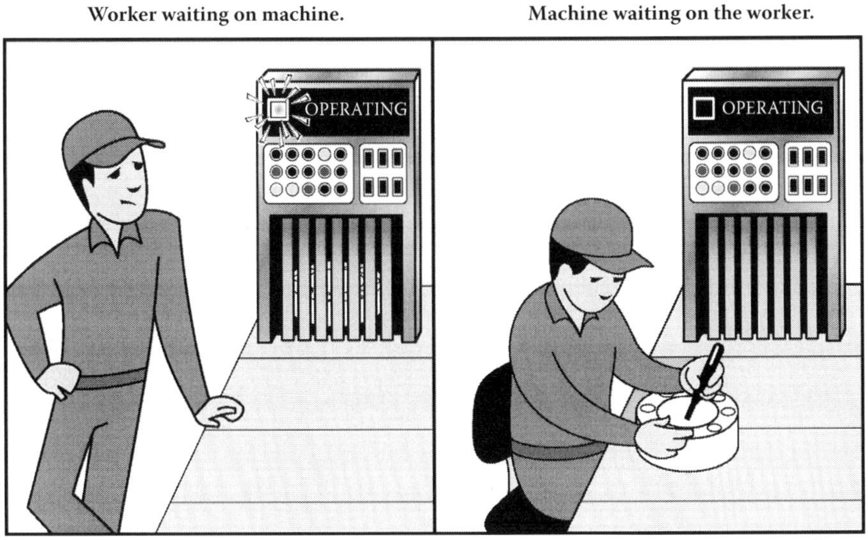

Worker waiting on machine.   Machine waiting on the worker.

Figure 5.9   Waiting.

process the required work within the takt time and the required production levels can be achieved during the normal shift, then I would say that the second scenario is the best way. That does not mean that it can't be better; however, if the machine has excess capacity, then it is better to fully utilize the worker. If the machine cycle time is equal to the takt time, this tells us that we may be capacity constrained. It is never good for the equipment cycle time to equal the takt time, unless it is an automated cell. If this is the situation, then I would say that the first scenario is the best way.

Most of the time, people have a tendency to think that waiting is good in terms of manufacturing, because generally waiting indicates that you are ahead of the production plan. If this is the case, then we can see that overproduction and overstock can cause waiting. Having more parts than necessary gives people a sense of safety, a comfort cushion. The paradox of this, however, is that most people cannot stand to wait and do nothing. So, what do they do? They fill the time they have for waiting by performing other work; this NVAW generally leads to overproduction. I think it is fascinating to consider how waste generates waste.

Let us assume that a machine breaks down. If there is any preparation work that needs to be done before the parts are to be machined, you can generally find the line workers filling that wait time by preparing those parts. That way, when the machine comes back on-line, they can hit the ground running with a faster pace. This is only one way in which overproduction can disguise itself as waiting.

These are only some of the things we look for as symptoms of waiting. Although the causes of waiting can vary to an infinite degree, the main reasons can usually be narrowed down to a few. Perhaps there is unevenness in the manufacturing cycle, or maybe the cycle time of the product does not match the takt time; it could be ill-conceived equipment or process layout. A common condition would be an imbalanced condition; on an assembly line there could be some process that is not balanced appropriately, and thus some workers are waiting while others are overburdened. Probably the most familiar cause of waiting is batch production. Batch production creates waiting time because each process is producing a specific number of pieces or products. So, if one process has a cycle time of thirty seconds, and the next process has a cycle time of four minutes, they would need very specific and appropriate controls to ensure they do not create unnecessary inventory.

The countermeasures vary for each of the problems just listed. For uneven flow, we would look at leveled production, heijunka, to improve our process. If an ill-suited equipment layout is causing an imbalance, we would look at creating a U-shaped equipment layout; this way, we can have one operator easily operating more than one piece of equipment. If there is a quality problem, try to install a poka-yoke device that will either detect or correct the problem before the defect occurs, or stop the line so the problem can be corrected prior to passing on the product to the next process or customer.

5.3.3 Overstock

Back on our assembly line, we can see another type of waste has manifested itself (Figure 5.10). In this example, overstock has manifested itself between operator one and two. As discussed earlier, this indicates a

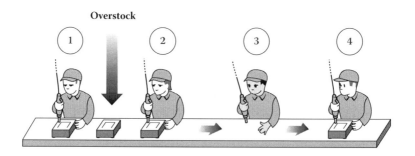

Figure 5.10   Overstock.

problem with process one or two. Even though overstock is a type of waste, it is often used to maintain a continuous flow.

Perhaps the easiest waste to observe, overstock in many respects is the hardest to eliminate. To eliminate or reduce overstock as much as possible, there are multiple countermeasures that can be applied. To apply them correctly, the cause and effect of each type must be quantified. Several of the causes of overstock can be traced to the manufacturing process and the organizational culture.

Overstock is creating more parts and components, or having more raw materials on hand, than are necessary to achieve the operating plan. Figure 5.11 is an example of how overstock can get out of control if the process is not managed.

Figure 5.11 shows a fine-looking warehouse, yes? Everything appears to be neat and easy to find; I bet if Toyota chronically overproduced, their warehouse would look just as orderly. Unfortunately, overstock such as pictured here will hide every type of waste.

If we look at the same image of a warehouse with the seven wastes in mind, what was once an orderly warehouse becomes one giant liability.

Generally, the effect of overstock is that a long lead time is necessary before any material becomes a finished product. The time it takes for a product to move from raw material to finished goods is referred to as

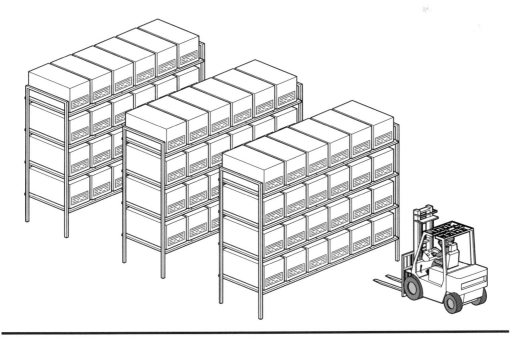

Figure 5.11   Orderly Warehouse.

throughput time. If I have a day's worth of overstock, then that stock is not going to become a finished product for a day. Now, if I have two years of overstock, which I have seen many times, it will be *two years* before I see a return on the investment made for the raw materials, not to mention all of the associated costs of labor hours, processing, and warehousing. I have seen many environments, especially in the automotive industry but also in many other manufacturing facilities, where the executives and management believe they have run out of space to contain their overstock. So they look at building a warehouse to alleviate their needs. It is only after eliminating inventory that the executives and management discover that they have more space than they previously thought. One of the biggest problems that contributes to overstock is the fact that most people, from the company president to the line worker, understand that stock is necessary for manufacturing. One thing that generally is neglected is understanding how to control the level of stock in the operation. We need to have parts and subcomponents to make a finished product. WIP is necessary to facilitate continuous flow. The problem is when you have WIP in your process that is days, weeks, months, even years old. To control the levels of stock in the operation, standardized work has to be in place. Standardized work is a countermeasure for reducing overstock in an operation. Of course standardized work has other benefits, but in reference to stock, it helps to define how much of what is necessary and when. Without standardized work in place, an organization will not be able to control the level of stock and maintain an optimum level of productivity.

Whenever I visit a facility I look at the levels of stock, both raw materials and WIP, in an operation as well as the level of FPI. The reason I do this is to determine the effectiveness of the organization's scheduling system. More often than not, the scheduling system creates WIP and FPI and fails to manage raw materials.

5.3.4 Overproduction

Overproduction is simply producing more finished products (FPI) than are necessary to fill the available orders (Figure 5.12). The outline that most managers and supervisors follow is based on a "feeling" that producing more than necessary is logical in case of machine breakdown or general downtime. This gives them a buffer, a comfort zone, if you will, that allows them to feel at ease, so that if "something" happens, they can still fill customer orders. There is nothing wrong with having a buffer as long as it is managed.

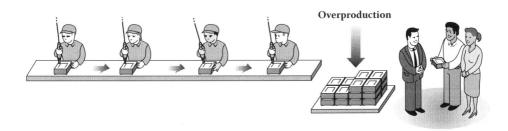

Figure 5.12 Overproduction.

I once was working with a large automaker that was laying out a new vehicle plant. As I reviewed the plant layouts, I saw that there was no buffer between the manufacturing operations. I asked why they had not put in a buffer, and they told me that a consultant who was an "expert" on the Toyota Production System told them that Toyota did not have buffers in their plants. They had literally set up the plant to achieve one-piece flow. I told them that they were nuts! In a factory as complicated as one that produces automobiles, it is essential that a buffer exist from one manufacturing area to the next. In Toyota, there are even buffers between the individual assembly lines. This again is an example of a misunderstanding in the manufacturing community that Toyota exists for the purpose of implementing the perfect production system. Toyota, like every good business, exists to make money. The key to making money in manufacturing is producing high quality with a reasonable cost.

Many people are also confused between overproduction and overstock. In manufacturing, there are four basic types of inventory in an operation:

1. Raw Material—materials or components that need to be manufactured to produce a value-added product
2. WIP (Work in Process)—product that is partially processed and is not in-process stock
3. In-Process Stock—product that is directly being manufactured in one of the manufacturing processes
4. FPI (Finished Product Inventory)—finished products that can be sold to a customer

A key to understanding overstock and overproduction is the "over." In a manufacturing operation, we are taking raw materials and processing them into a product that has value for the customer. Of course you are going to have some level of inventory in all four categories. From a waste standpoint,

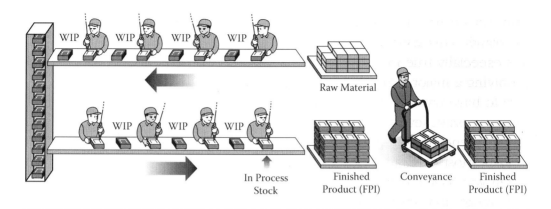

Figure 5.13 Inventory Examples.

anything that is in excess of what is necessary to efficiently manufacture the product is in the "over" category.

Looking at Figure 5.13, we can see that overstock can manifest itself in three of the four inventory categories; raw materials, WIP, and in-process stock. Overproduction applies only to finished goods.

Overproduction also can make managing an operation more challenging. Should there be a quality problem with the finished products, the cost for sorting and repairing the products increases. This could affect the level of quality that is delivered to the customer. Overproduction also hides manufacturing problems. Good managers want to see problems, not hide them. Problems are easier to fix once they are visible.

One of the goals of the Toyota Production System is to produce products in the quantity needed by the customer. Only produce what the customer is willing to purchase. Products only have value when there is someone who is going to purchase them.

I once visited a glass manufacturing plant, and there was inventory in all its forms everywhere you looked. As we were touring the plant there was literally months' worth of inventory at each stage of the manufacturing process. When we finished the plant tour, the director of operations said, "Would you like to see the distribution center?" I thought to myself, how could there be more? When we arrived at the distribution center, it was a massive 1.2 million square foot facility completely full of finished goods. Some of the finished goods were over six years old. This is a classic example of producing more products than the customer is willing to purchase.

In Toyota, this would not be possible based on the manufacturing controls that are in place. Another big difference is that Toyota supplies their own retail channel, which gives them control over the level of orders to the plants. This is one of the biggest challenges in manufacturing today. It is especially true in the retail industry. When you are a small company supplying a major retailer like Walmart or Target, there is a lot of pressure to have product on hand when it is ordered. This is not an excuse to carry excessive levels of FPI, but it does present some challenges. Another key difference between Toyota and most companies is that Toyota is self-funded. Most companies today operate with some sort of lending facility. When you have a capital-based lending facility and you want to begin the implementation of some of the concepts discussed in this book, it may seem counterintuitive. For example, if your bank has allowed you to borrow money on your inventory and then you reduce your inventory, you could cause some problems with your liquidity. This is a challenge that Toyota is not faced with. Of course, we all wish that we did not have the banks breathing down our necks and that we were self-funded, but for most of us this is not the case.

I think it makes sense to everyone to not have excessive levels of finished goods. We would all be very happy if we could sell all of the finished goods and draw the level of inventory down immediately. Again, this is the perfect-world scenario. I once had a discussion with a Toyota "expert" from the consulting world, and we were discussing with the CEO of a company how to reduce the levels of finished product. His suggestion was to shut the plant down for two weeks and draw down the level of inventory. This seems like a good idea on the surface, but let's make some basic assumptions in our example:

Finished product inventory value = $5 million
Inventory borrowing base rate = 0.50
Fixed cost rate = 0.60
Variable cost rate = 0.40
Weekly costs = $3 million

In this example, the bank allows us to borrow fifty cents on the dollar based on our capital-based lending facility, which means we are already using two and a half million dollars of the value of the inventory to fund our company. Our fixed cost structure is 60% of our total costs and making our variable costs 40%. For the sake of this example, let's assume that the five million dollars of inventory is wanted by the customer. If we shut down for two weeks and reduce our FPI to zero, what have we achieved? If we sell the entire inventory, that generates five million dollars of cash, of which we have to pay two and a half million dollars to the bank because we were already borrowing fifty cents on the dollar. During the two weeks of shutdown, we still had to lay off all of our hourly workers and did not pay them, and we did not have any of our variable costs. However, we did still have all of our fixed costs and that equals more than three and a half million dollars of cost for the two-week period. Let's do the math.

Sell all inventory	$5 million
Repay banks	$2.5 million
Fixed costs	$3.6 million
Value to company	($1.1 million)

What? We removed five million dollars worth of inventory, and it costs the company money? This is one of the major differences in a self-funded organization and one that relies on a lending facility to operate.

Although the example points out some of the challenges of implementing TPS in the real world, it is not an excuse to have high levels of inventory. It just illustrates that everything is not as simple as what is written in a book. How do we reduce inventory and not have a negative effect on the company? Hopefully you will find those answers and more as you read this book.

5.3.5 Repair

Repair is perhaps the most obvious waste to spot, and if approached correctly, also one of the easiest to remedy. However, it seems as if most companies are content to focus on repairing defects, as opposed to actually preventing them from occurring in the first place. The cumulative effects of defects will be an increase in the cost of the product due to decreases in productivity levels, which increases the total man-hours due to the need for

Figure 5.14   Repair Process.

inspection. Depending on your customer requirements, you may be forced to hire an outside company to inspect and sort the product prior to shipping it to the customer. For those of us who have worked in the auto industry, we understand how quickly those costs can impact the bottom line, not to mention the potential damage to the reputation of the organization. This could lead to more difficulty in the future to win new business. As a business, our goal is to make money, and controlling defects *saves us money,* which in turn creates stability for us, our employees, and our investors. I often tell the management teams in our companies that we work really hard to make money; we should work equally as hard to keep some of it!

The need for repair is caused by defects in the products. If the products are manufactured correctly, then the need for repair is reduced. For example, if I am processing metal that has to reach a certain temperature before I can put it through the die cast process, and it does not achieve that correct temperature, the likelihood that I am going to create a defect is increased. It is that simple. Sometimes it is as basic as being aware of the process control parameters in order to avoid creating defective products in the first place.

Many times I have observed instances where inspection tolerances were so tight that parts within standards were labeled as defective, thus increasing the level of products needing repair. As a result, inventory levels are increased and manufacturing costs are increased. Having an understanding of what is good, what is not good, and the parameters of what makes a process stable and repeatable will allow us to limit the need for any additional

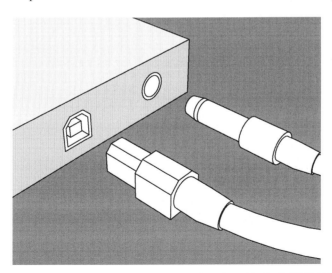

Figure 5.15   Poka-Yoke Example.

manpower, as well as create an environment where we can have standardized work, which results in lower manufacturing costs.

There are a few anecdotes in the history of lean that have attained mythical status, and the best known is how Shigeo Shingo opted for the term *poka-yoke* (mistake-proof), as opposed to *baka-yoke* (fool-proof). The story goes as such: Dr. Shingo was addressing a group of part-time workers at an automobile factory. His topic of discussion was how to make the process of spot-welding seat frames as fool-proof as possible with the introduction of a baka-yoke device. Upon hearing this, the part-time worker primarily responsible for the operation burst into tears, thinking that she was considered a fool by Dr. Shingo, as well as by her colleagues. From that moment on, Dr. Shingo coined the term *poka-yoke* to avoid any implication that the devices were needed because the workers were fools.

Poka-yoke as a countermeasure is nearly unbeatable. It can be applied to nearly every aspect of production: equipment, parts, materials, and, more importantly, the process itself. When any poka-yoke device detects an error or abnormality, it will trigger either the machine or the entire production line to stop. Before the line or machine starts up again, the defect issue will have been resolved. In this manner, quality is built into the process, which is the most effective way to eliminate waste. This concept is called *jidoka* inside of Toyota, and it is one of the two pillars of the Toyota Production System.

It does not matter what type of product that you are manufacturing; if a customer buys a defective product, you have created waste. One thing that many companies seem to have forgotten is that, more importantly than waste, you have created an unhappy customer.

5.3.6 Overprocessing

Overprocessing is when more work is performed than necessary to process the work. Overprocessing is often difficult to identify in a facility where you are familiar with the process. Many times people who are close to the process have become accustomed to it and will classify overprocessing as VAW. In the production line example in Figure 5.16, the operator in position one is hand-starting a fastener and then the operator in position two is tightening the fastener. Hand-starting the fastener does not add value to the product; only the actual tightening adds value. To identify overprocessing, we have to have a good understanding of VAW.

Another way of looking at overprocessing is any extra step in a process that adds cost but no value. Easy, right? For instance, let us assume that we

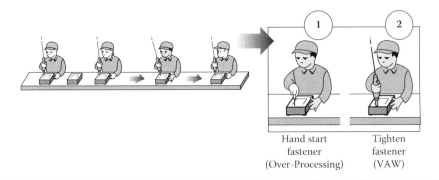

Figure 5.16 Overprocessing.

have to attach a label to a particular part before it can be considered complete (Figure 5.17). The first thing that has to be done is to peel the backing off the label to expose the adhesive side. The motion of taking off that label backing is NVAW; it is overprocessing because it is not necessary. Now, of course the backing has to be removed for the label to stick, but is it something that we have to do? There are label guns that remove backings as they apply, as well as dispensers that remove backings as they are pulled. If we do not properly classify waste, then we will more than likely not be able to implement the appropriate countermeasure.

If you are thinking to yourself that something as simple as removing the backing from a label is a trivial improvement at best, then I do not know if

Figure 5.17 Overprocessing Example (Label Installation).

there is anything useful you can learn from this book. For example, let's say you own one manufacturing facility that only has four production lines, and each line has an operator who labels parts or products as part of his or her job function. The physical act of grabbing a sticker, removing *and* discarding the backing, and then applying the sticker in the appropriate place and position takes five seconds. The operator repeats this process 40 times an hour. This equals to 3.5 minutes per hour being devoted to applying a sticker. Big deal, right? After an 8-hour shift, that number is over 26 minutes; after a work week, 2 hours and 13 minutes; after a fiscal quarter, 27 hours. So, 27 hours per quarter for one worker to apply labels; and we have four production lines, each with a label process. All added up, by the end of the fiscal year, you have paid out 432 man-hours for applying a label.

Now, let us say that you have introduced a simple label gun; when you squeeze the handle, the sticker comes out with the adhesive backing exposed (Figure 5.18). Another example would be to have the labels manufactured onto a roll and then develop a simple jig that allows for the labels to be removed without removing the backing paper from each individual label. With either kaizen, the label installation process now only takes 2 seconds, or 1 minute and 20 seconds per hour. This adds up to roughly 10 minutes per 8-hour shift; 50 minutes a week; 3.3 hours per month. The fiscal year total for all four production lines is 120 man-hours, a savings of 312 man-hours per year. The label gun and jig have saved you quite a bit of money for such a trivial process. Not bad, right? Truly, it is not bad considering that the gun and jig are a rather passive countermeasure. A more active,

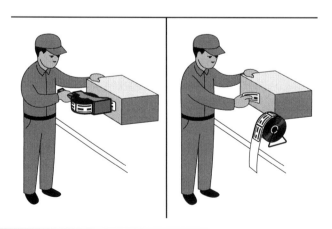

Figure 5.18 Overprocessing Kaizen Examples.

dynamic improvement would be to automate the label process and take such a menial task out of the hands of your workers.

The effects of overprocessing are found in the abundance of operators and processes needed for production. Accordingly, as quite a bit of overprocessing comes from bad process sequence and bad work sequence, productivity will decline due to an increase in repairs from those bad sequences. Remember that repairing defects is 100% overprocessing. The other suspects, bad flow, tools, jigs, and the lack of standardized work, are all here as well. Some of the most effective countermeasures here are to do cycle time balances, rebalance the workload, and build-in quality to the process. Approach all improvements from the standpoint of common sense and repeatability.

5.3.7 Non-Value-Added Work (NVAW)

Many people really get confused about NVAW; after all, isn't all waste non-value-added? Absolutely, all waste is non-value-added work. In Figure 5.19, we can see that the worker at the beginning of the line is placing a worker order manifest on the line to tell the other workers what type of product they are making. Although this form is useful in the operation, it does not add any value to the product. This differs from the example we used to explain overprocessing with the label application in that the application of the label is VAW. The application of the manifest does not contribute to the operation other than to give instruction. Another good example of NVAW is removing components from the boxes; the removal process adds no value, only the installation. This too is not overprocessing, because the removal of the component refers only to the component and not to the end product.

Once again, like overprocessing, transportation, and waiting, NVAW is hard to identify because it hides so well among other wastes, as well as in the perception that it, too, is necessary. NVAW work is any work that does not add value, in function or appearance, for the customer. For all of us who

Figure 5.19   Non-Value-Added Work.

make our living without materially contributing to the production process, I am sorry to say that we are a classic example of NVAW. For any executive who might be reading this book, if you can reduce members of the management team, you are eliminating waste! However, please do not eliminate them until after they have read my book.

All joking aside, if your process is filled with NVAW, you will have workers who are very busy but are contributing very little value, if any, to the final product. NVAW creates an unstable work environment by introducing fluctuation into the manufacturing process, which negatively affects quality and productivity. Thankfully I can say that most NVAW is actually unnecessary and can be easily eliminated from the process.

When I am visiting a company for the first time, I generally like to spend time with the management team to understand the process. Once I have a basic overview of the process, I like to visit the plant floor and walk the operation from the start of the process through the final processes of the operation. During these visits, I am observing the process and making a judgment concerning the effectiveness of the current state operation. As a part of this process, I need to make a high-level assessment to understand what level of improvement can be made in the process. Understanding the level of NVAW is a key component to help me understand the overall efficiency of the process.

I generally have one to two hours to walk the floor in order to make a high-level estimate of the opportunity. I employ a process that I was taught at Toyota in Japan, called *teashi*. The literal translation means hands and feet. By observing the hands and feet of the workers, I can determine the general level of productivity in the operation (Figure 5.20). It is not possible for a human to add value to a product without using their hands and their feet. When I am looking at the hands of the operators, I am making general observations of the percentage of time that their hands are idle, or are performing NVAW. Observing the feet of the worker is a relatively simple method for understanding the percentage of time in the process that the worker is walking. By observing the hands and feet together, I can determine if the worker is working while walking, or just walking.

Using this simple process, I can generally determine the improvement opportunity of an operation within 10% of the actual opportunity.

Understanding NVAW is essential for understanding production efficiency. Even though NVAW is the most apparent type of waste in a manufacturing or production process, it is the most difficult to identify and to countermeasure.

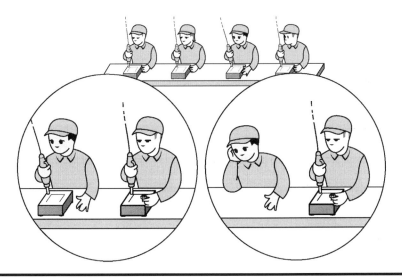

Figure 5.20   NVAW Observation.

As you hone your skills, you will understand that identifying and correctly classifying waste is a foundational element of making real improvement in any business process.

Many people will tell you that elimination of waste is the key; however, I have found that even someone without an understanding of the Toyota Production System will eliminate waste when the waste is apparent. The real challenge is to identify and correctly classify the waste.

5.4 Muda Countermeasure Methods

Now that there is a basic understanding of the seven types of waste, we can begin to look forward toward elimination of waste from our process.

The muda summary chart (Figure 5.21) is a great tool for identifying and classifying waste. On the left are the seven types of waste. As you move across the page, the next two columns summarize the difficulty for identifying and then eliminating the waste. The next column indicates the reaction that management should take once this type of waste has been identified. Some people may tell you that you have to eliminate all types of waste; however, like anything in management, it is necessary to prioritize the opportunities. The reaction indicated for each one is just a suggestion; you will have to determine the priority based on your circumstances. The next column lists some of the common ways to identify the type of waste. This is not meant to be a checklist but just a helpful illustration to properly classify

MUDA (7 Types of Waste)	Identify		Mgmt Reaction	How to Identify?	What are the causes?	How to Fix?
	Identify	Eliminate				
1. Transportation - moving products from place to place creates no value	Easy	Difficult	Expose and minimize	• Space - multiple stocking locations • Overstock • Increase in transportation labor • Increase in transportation equipment • Decreased efficiency • High levels of NVAW • Multiple touches of product	• Bad layout • Lot production • Inflexible machines • Poor standardized work • Poor scheduling	• U-Shaped layout; flow production • Heijunka - leveled production; rapid die change • Multi-functional equipment • Implement standardized work • Kanban system - PULL system instead of PUSH
2. Waiting - overstock hides waiting	Moderate	Moderate	Expose and eliminate	• Operator, T/M, machine etc. waiting. • Overstock- NOTE: most of the time waiting is not visible because people tend to fill up the time by doing NVAW or by overproducing	• Unsmooth Flow • Bad equipment / process layout • Inconsistent production flow • Unbalanced processes • Lot production • Poor standardized work	• Heijunka production • U-Shaped layout; flow production • Implement pace setters; andon system • Process planning • Heijunka - leveled production; rapid die change • Implement standardized work
3. Overstock - any stock that is not moving	Easy	Easy	Eliminate immediately	• Long lead time to become finished product - many components waiting to be processed • Space - takes up more space than required • Increased inspection and transportation	• Misconception by everyone that stock is needed • Bad equipment / process layout • Lot production	• Training of management; value stream map • U-Shaped layout • Heijunka - leveled production; rapid die change • Continuous flow

Figure 5.21 Muda Summary Chart.

	Identify		Mgmt Reaction	How to Identify?	What are the causes?	How to Fix?
	Identify	Eliminate				
MUDA (7 Types of Waste) 4. Overproduction - producing WIP and FPI that are not needed	Easy		Eliminate immediately	• Increased operating costs - components are paid for and not converted to finished goods • Increased working capital - operating costs are inflated • Increased labor $ - labor is higher than required • Overproduction creates overstock • No flow - push system • Quality is difficult to control • Multiple storage locations • Overstock - overproduction causes overstock • Reduced flexibility	• Unsmooth flow • Poor scheduling • Poor material handling methods • Poor standardized work • Lot production • Overcapacity • Too much labor • No pace setter for the operation • Poor scheduling • Poor standardized work • Lack of Takt time	• Rapid die change • Kanban system - pull system • Implement standardized work • Heijunka - leveled production; rapid die change • Scheduling and resource planning • Labor planning • Implement pace setters • Kanban system - PULL system instead of PUSH • Implement standardized work • Develop Takt time for each product line
5. Repair - All defective products are waste	Easy	Moderate	Monitor and eliminate	• Fluctuation in material and component cost • Low efficiency • Increased inspection processes • Increased labor	• Defective vendor parts • No in-process inspection methods or processes • Incorrect inspection method / standard	• Implement receiving Inspection system • Poka-yoke; In-process quality • Confirm quality expectations; standard

(Continued)

Figure 5.21 (Continued)

MUDA (7 Types of Waste)	Identify	Eliminate	Mgmt Reaction	How to Identify?	What are the causes?	How to Fix?
(continued)					• Process quality not equal to customer expectation • Poor standardized work	• Confirm customer expectation and process capability match • Implement standardized work
6. Over-processing - any step that does not complete an operation	Moderate	Difficult	Expose and minimize	• Prep work or work that does not complete a process • Increased NVAW rate • Reduction in productivity • Poor quality • Multiple touches of product • Rework and repair	• Poor process sequence • Poor work sequence • Poor tools and/or jigs • Not doing standardized work or standardized work is complicated • Defective vendor parts	• Re-sequence process • Implement standardized work • New tools and jigs • Process planning • Implement receiving inspection system
7. Non-value added work	Easy	Difficult	Expose and minimize	• Low efficiency • Increased labor • Fluctuation in process cycle times • Not working to completion, partial process	• Bad layout • Lot Production • Inflexible machines • Poor standardized work • Poor scheduling • Unsmooth flow	• U-Shaped layout; flow production • Heijunka - leveled production; rapid die change • Multi-functional equipment • Implement standardized work • Kanban system - PULL system instead of PUSH • Heijunka production

Figure 5.21 (Continued)

waste. The next two columns are tied to one another; the first of the two illustrates some of the general causes for that type of waste, and the corresponding row in the next column indicates the appropriate countermeasure for that particular cause.

Like all of the tools used in the Toyota Production System, the muda summary chart is just a tool, and the success or failure that you will have with correctly identifying and eliminating waste is completely decided by your execution.

I would give Toyota credit for these wastes, but chances are you have them too. It is most important to remember that there is no correct answer that can be applied to all situations. I have given suggestions for appropriate countermeasures, but they are only suggestions. The Toyota Production System is about achieving the best condition, and each company's best condition is unique to that organization.

5.5 Waste Elimination Example

Now that we have established a basic understanding of the principles of the three M's and the seven wastes, I would like to briefly illustrate some of the basic principles that I have developed to improve, or kaizen, the production process. These principles are not reference edition principles that can only be effective in a utopian organizational environment; they have been developed based on my twenty-two years of operational experience. These principles have been applied in a wide degree of organizations within and outside of the Toyota family of companies.

One of the areas where I was able to develop a certain level of proficiency at Toyota was the ability to go into failed business units inside the organization and turn them around quickly. To consistently get successful results in a time frame that would impact the operation, I developed a systematic process for looking at the operation, to which I applied all of the principles from my education at Toyota. The first step of my kaizen process begins with the principle of *genchi genbutsu*. At the end of Chapter 4, I introduced genchi genbutsu as an action-oriented principle for managing any operation. Genchi genbutsu is essential for any manager who wants to drive improvement in the organization. The whole premise of genchi genbutsu is that of execution. As any good manager knows, any plan that has ever been developed is only as good as the organization's ability to execute the plan.

To practice genchi genbutsu, it is essential for operational managers to spend time in the work environment with the people doing the work. This is one of the most basic concepts, and yet it is the one concept that many managers fail to understand. When I visit a company and all of the operational managers have offices separate from the production facility, without fail the production operations have glaring opportunities for improvement. How can a supervisor manage people if he spends no time with the people he is responsible for managing? Organizational execution happens in real time and must be managed by real-time managers. Managers who rely on data collected at the end of the day, week, or month are not effectively managing the organization.

Now that we are managing our people where they are doing the work, it will become evident that there are opportunities for improving the process. It does not matter how we define *work*. The fact is that, for us to execute the principles of TPS, we have to spend time with the people adding value to the process. Remember that management is a form of NVAW.

Now that an area for improvement has been selected for improvement, or kaizen, what do we do? How do we actually improve the process?

The key to making improvement in any process is to correctly classify the work. Earlier we discussed the difference between NVAW and VAW. VAW is only the element of the process that the customer is willing to pay for. For that reason, when we classify work, it is important that we have a very clear understanding of what kind of work it is, and what it is not. When I am looking at a process, I classify work into three distinct groups:

1. Value-added work
2. Unnecessary non-value-added work
3. Necessary non-value-added work

The key distinction is correctly classifying unnecessary NVAW and necessary NVAW. Although all of the work is non-value-added, the approach for improving the process significantly changes based on this distinction.

As we look at a process for a kaizen event, we need to be able to identify work from waste, and then differentiate work from NVAW. Within the boundaries of NVAW, we need to understand the differences between necessary and unnecessary NVAW.

Unnecessary NVAW is relatively simple to countermeasure, as opposed to necessary NVAW and VAW, which are much more complex. When initiating a process kaizen event, I always emphasize to the team that they should

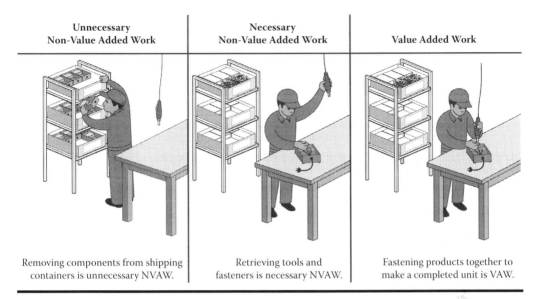

Unnecessary Non-Value Added Work	Necessary Non-Value Added Work	Value Added Work
Removing components from shipping containers is unnecessary NVAW.	Retrieving tools and fasteners is necessary NVAW.	Fastening products together to make a completed unit is VAW.

Figure 5.22 Types of Work.

focus the improvement activity on the aspects of the process where they have direct control. This ensures that the team does not become frustrated when identified actions rely on areas outside the scope of the team. This also ensures that tangible benefits are extracted from the event.

Looking at Figure 5.22, unnecessary NVAW is work that is completed but is not necessarily needed to be completed in order to make a completed product, or unit. In this example, the worker is removing the component from the shipping container. Some may classify this as necessary NVAW, but this is not correct. It may be necessary that this product needs to be shipped in this particular container from the vendor to ensure quality, but for this process, this work is not necessary. This work is only required because this is how the part is presented to the worker. When we are doing a process kaizen, we have to look at the process and see how we could optimize the process. Therefore, if the part presentation were modified, the amount of NVAW could be reduced or eliminated.

In Figure 5.23, the unnecessary NVAW is eliminated, as the component is now placed on a conveyor. The part is presented to the worker in a way that the time required for removing the component from the box has been eliminated. Assuming the shipping container from the vendor does not change, the component has to be removed from the container by someone, so where is the actual process improvement? The improvement comes from two sources. First, the worker now has the part presented on the work surface and does not need to turn around to remove the component. This not only

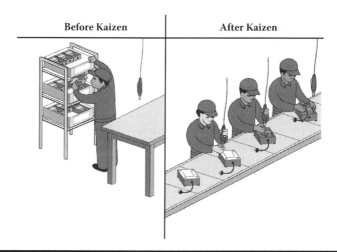

Figure 5.23   Unnecessary Non-Value-Added Work Countermeasure.

reduces the time that was spent removing the component, it also reduces the time to turn around and walk to the storage location. Even if we assume that someone still needs to remove the product and place it on the conveyor and that time is equal to the time the worker spent removing the component from the container, the benefit is the time to turn and move to the container storage location. Even though the savings is small, the savings would never have occurred if the work had not been classified correctly. At Toyota, we would analyze a process to save a half second from a process with a cycle time of fifty-five seconds!

If we look back at Figure 5.22, we can also see that the necessary NVAW has been identified as the work required grasping the air tool and picking up the fastener. The process to countermeasure the necessary NVAW differs greatly because without changing the components or the final product, we cannot eliminate the time necessary to complete these steps. The component is designed so that the two components have to be assembled together to make a completed unit. Without changing this design, we cannot eliminate this time. Again, the proper classification of this type of work is essential. Even though this necessary NVAW couldn't be improved in the current process kaizen event, we can provide this information to the engineering group for consideration when a design change is necessary.

Finally, the VAW illustration in Figure 5.22 shows the actual part of the process that is value added. Of this process, only the actual tightening of the components is classified as VAW. Because the two components are purchased from vendors, the value that the worker provides for the customer is by assembling the two components together to form one product. The individual components have no value to the customer because the customer is only willing to purchase the finished product. Often when I am discussing VAW and NVAW with management, they assume that the VAW rate of the product is the majority when, as this example illustrates, it is usually the opposite. The first time we did a thorough analysis of the assembly processes at Toyota's facility in Georgetown, Kentucky, the VAW rate was 27%! Don't be shocked if the VAW rate of your process is much lower than this.

Because VAW generally entails a design change to parts or components to accomplish, again, this is something that is very difficult to improve during a process kaizen event. Given this illustration, it is essential to classify the type of work properly to determine where the real opportunity for improvement is in the process.

Chapter 6

The Golden Rules of the Toyota Production System

6.1 Fundamentals

To apply the principles that we have discussed in the first five chapters, we also have to understand some basic fundamental principles. Fundamentals are important when trying to create an action plan. Without understanding the fundamentals, execution suffers. We can see this concept repeated in the world that we live in every day. How many times have we heard a football coach talk after a loss that the team needs to focus on the fundamentals? Another practical example of this can be seen in one of the biggest challenges a parent faces: teaching a teenage son or daughter how to drive.

Automobiles today are complicated machines with miles and miles of electronic wires working with the engine of the vehicle to make it function based on the instructions received from the operator (Figure 6.1). Even though the operator does not understand the details of how the internal combustion engine works, once the operator has a basic understanding of the fundamental principles—steering, braking, and accelerating—he is able to effectively operate the vehicle. Once the operator has mastered these basic principles, there are certain elements of the operation that have to be monitored to make sure the automobile functions as intended.

Today automobiles have systems that monitor these functions, and the operator needs only to respond to the warnings provided by the vehicle (Figure 6.2). If the vehicle experiences a problem, an indicator will light up in the instrument panel telling the operator that there is a problem. If the operator has the skill set to fix the problem, he or she will complete the

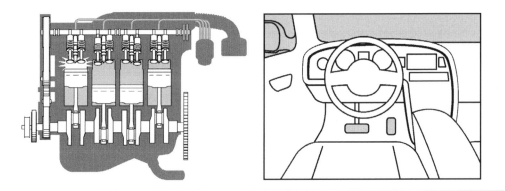

Figure 6.1 Automobile Function Example.

repair and restore the operation of the vehicle to normal. If the operator does not have the expertise to solve the problem, the vehicle is taken to a specialist to diagnose and repair the problem.

This same process applies to understanding the Toyota Production System. When an organization is attempting to implement the TPS, it is essential to understand that although everyone needs to have some basic understanding of the system, it is not necessary for everyone in the company to have the same level of understanding. There are certain foundational elements that all members of the organization must understand. These foundational elements, or principles, are what I call the golden rules of TPS.

Going back to the example of teaching a teenager how to drive, it is not essential that he be tought how the car works. He doesn't need to have a complete understanding of the mechanical and electrical systems of the vehicle. It is only essential that he understands how to operate the vehicle and where to take the vehicle when it is not acting as intended so any problems can be resolved.

Figure 6.2 Automobile Warning Light and Mechanic.

This same philosophy is applied at Toyota with the understanding of the TPS. Although everyone at Toyota interacts and is a part of the production system, it is not necessary for everyone to be an expert in all aspects of TPS.

At Toyota, we spent a lot of time determining the fundamental skills necessary for the line workers and training them on these fundamental skills. Many of the tools used to implement the TPS are not completely understood by all of the workers, but because the workers understand the fundamental principles, they are able to support the implementation process.

6.2 The Golden Rules of TPS

There are many ideas and visions for implementing the TPS. Much of the information available today focuses on the tools of TPS and not on the principles. The material available concerning principles focuses on philosophical principles, not real-life principles that can be defined and implemented. These three principles have guided my understanding of the implementation of the TPS for over twenty years.

Simplify
Standardize
Specialize

These principles can also be referred to as the three S's, but this can be confusing, especially when discussing the 5 S's; therefore, I simply refer to these as the golden rules.

6.2.1 Simplify

Simplify means exactly what you are thinking. The basic principle is that whatever we do should be so simple that someone walking off the street should be able to understand what we are doing and why we are doing it.

From my experience working with various manufacturing and operational companies, I see a pattern that exists, where many companies overcomplicate their products and processes. Often when I am meeting with a CEO or senior operations person in a company, the first thing they do is explain how unique and complicated their processes are in comparison to a competitor. Many times before I visit a facility, people will check my background and see that I worked for many years with Toyota, and they will tell me that

Which is more complicated? Hmm...

Figure 6.3   Simple Image?

they know I have a lot of experience in manufacturing, but their widget is more complex to manufacture than most other components. I even had one CEO who was producing a very simple electrical component inform me that the process of manufacturing the component was much more complicated than, say, producing an automobile! I want to tell them that the problem they should be solving is why they have developed a complicated process to manufacture a simple product.

In Toyota we say that we should make everything so simple that even a monkey could understand the process (Figure 6.4). Developing a simple process sounds so … simple. Actually most companies have the capability to manufacture their products; the real challenge is to find a simple method for producing even the most complex products.

Figure 6.4   Monkey on Production Line.

I remember when I was being taught, by one of my teachers, some fairly complex thought processes relating to the TPS. He was teaching me how to develop strategy documents to explain and justify the projects we were undertaking in the vehicle assembly plant. The typical assembly manufacturing process usually involves well over a thousand operators on the shop floor and literally hundreds of various types of automated and manual equipment processes. We would spend days developing a single strategy document. My teacher was always sending me back again and again and again to make the strategies simpler. We were developing very complex ideas that would entail the company spending millions of dollars and sometimes would determine our direction for the next three to five years. One of the most challenging points was that I was always forced to contain my strategy on a simple one-page 8.5 × 11 sheet of paper. Oh how I longed to use the now famous A3, or 11 × 17 paper. I did not fully understand it at the time, but the one sheet of 8.5 × 11 paper forced me to simplify my strategy by using graphics and images instead of words to depict the current and future states as well as the details of the implementation process. I worked so hard at this that I soon became quite famous for my very simple strategy documents. It would not surprise me to see some of these same strategy documents being used today. The point of developing such simple documents was that not everyone in Toyota had the same level of understanding of the concepts that were being developed. By making the document very simple, we were able to clearly convey our ideas to the senior management in the organization to gain their approval, and then use the same document to explain the process to workers on the production line. The easy way is to create a PowerPoint presentation with a hundred slides; however, by containing the strategy to one piece of paper, anyone could pick up the strategy and understand exactly the targets and intent of the process.

The same principles apply to any manufacturing process. The key point is the ability to take a complex operation and simplify it so almost anyone can do it with a minimal amount of training. One of the concepts for simplifying the manufacturing operation is visual control. This is sometimes referred to as the visual factory. The point is that anyone walking on the production floor should be able to understand the flow of the production process.

Toyota is famous worldwide for its many methods of visual control and visual management. One of the most obvious things that people notice when they visit a Toyota facility is the visual control. Whether it is the 1500 mm standard for the storage rack height in the assembly plant or the

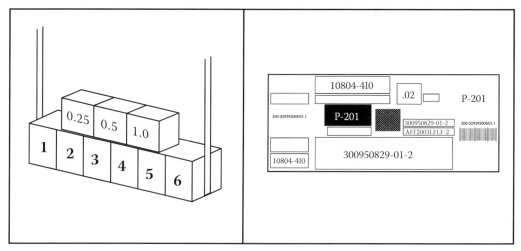

Andon Board   **Kanban**

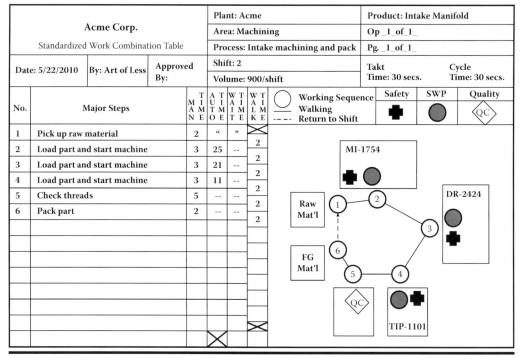

Figure 6.5 Simple Visual Control Examples.

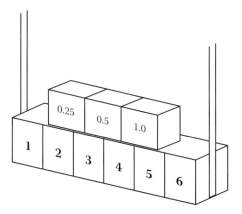

Figure 6.6   Andon.

andon systems that provide visual information for the flow of production, the visual control standard at Toyota is striking when entering a facility for the first time.

One example that is obvious when you enter the factory is the andon board (Figure 6.6). This is simply a production board that displays the pertinent production information during the manufacturing process. This is one of the fundamental systems that is understood by all of the workers on the production line. This simple board keeps the workforce informed of the production condition. On this board, if the line is running normally, the name of the line is lit up in green. If the line is stopped, the name of the line is not lit. If the line is waiting for work from the previous process, this condition is known as short and is indicated by the SHO on the andon board. If the process is waiting on the next process, the condition is known as full and is indicated by the FUL on the andon board. Various work positions are indicated in the andon board. These are activated by the operators and signal the supervisors where a problem is being experienced on the line. Several other factors such as quality, production targets and actual, and safety items are also displayed. Each production line has an andon board, and various summary lines are strategically placed throughout the operations so that the workers and the supervisors can monitor production and more importantly, respond when an abnormality occurs.

Another example is what is referred to as the key production indicator (KPI) board (Figure 6.7). On this board, all of the key performance indicators are displayed and updated on a daily basis. The items that meet the target are displayed in green, and the items that do not meet the target are displayed in red. This is where the management team will come together

Figure 6.7   KPI Board.

to discuss the problems identified and the solutions to those problems. Examples such as these enable the management team to quickly grasp the situation and to know where they need to go to improve the operation. The process of simplifying these processes highlights the significant information and minimizes management noise. This concept enables the management team to be much more effective in their daily management of the production floor. Even someone who has no real experience in manufacturing can attend the meeting and know the areas that need to be addressed.

Standardized work is another example of visual control that must be fundamentally understood throughout the organization (Figure 6.8). When we think of the complexity of the manufacturing process, it seems like common sense that workers need simple instructions that detail how to build the product. Over and over again, I experience organizations that have no systematic method for transferring the necessary knowledge to new employees. In the automobile manufacturing process, instruction is

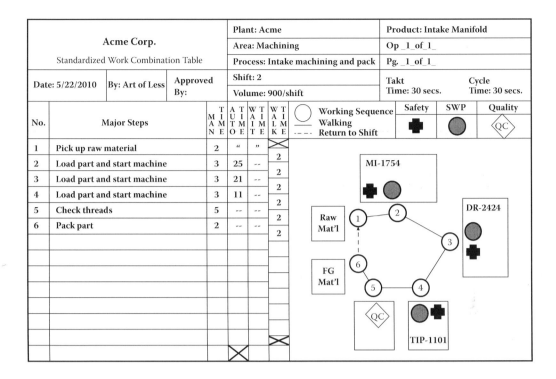

Figure 6.8   Standardized Work (STW) Chart.

needed for each operator on the production line to teach him which parts go on which vehicle. In Toyota, this process is especially complex because multiple vehicles are produced on the same production line. Workers may build up to five different models on the same line with countless variations to each model. This has become more pressing as Toyota expands their current product offering. As Toyota expanded the model lineup to capture additional sales, existing production capacity was retooled to produce the new models on the old lines. In one instance, the luxury sport coupe from Lexus was being produced on the same production line as a taxicab. One can imagine that there is not a lot in common between a taxicab and the Lexus SC470!

Recently, Toyota has spent a great deal of engineering resources to simplify the production methodology. This system of manufacturing focused on sequencing each vehicle's parts and components and delivering them just in time for the assembly of that particular vehicle. As one can imagine, this was a staggering undertaking. The method of implementation was a concept known in Toyota as set part supply (SPS) (Figure 6.9). This

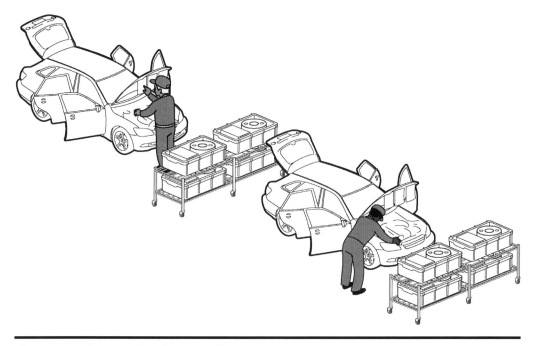

Figure 6.9   Set Part Supply Example.

system enables the workers to simply focus on assembling the vehicle while other workers select the appropriate parts and sequence with the vehicles. The goal of this process was to simplify the operations. So fundamental is the philosophy of simplification that Toyota literally has spent millions of dollars to retool factories around the world to incorporate this new manufacturing methodology.

The simplify concept also can be seen in the methodology Toyota uses for implementing automation (Figure 6.10). We would think that Toyota, with its dominant position in manufacturing and sales, would employ all of the latest technologies in the manufacturing process. Although this is true for many operations, it is surprising to see all of the manual operations inside the factory. In some factories, the process is entirely manual. This is especially true in developing countries like India and China, but many of the same manual operations can be seen at facilities in the United States as well. One obvious reason for this lack of automation is obviously to control investment; however, the main reason is to keep the manufacturing process simple. Although automation can make the process easier for the line worker, it does not necessarily make the process any simpler. In fact, the more automation there is, the more special maintenance and engineering resources are needed to maintain the equipment.

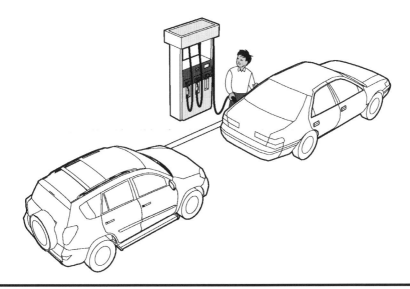

Figure 6.10 Automation Example.

6.2.2 Standardize

The next S of the three S's is standardize. Toyota is world renowned for the development of standardized work. Many people misunderstand the purpose of standardized work. In manufacturing, there can be numerous variables in the process; however, this is also true for non-manufacturing operations. As customers demand more diversity and customization in the products and services offered by companies today, the overall process of providing these products and services is becoming more complex. This complexity increases the number of variables, and these variables cause variation in the process that can lead to abnormalities. Abnormalities will result in poor efficiency and poor quality. Standardized work is a method to achieve repeatability. Any person who has worked in a manufacturing process will tell you that achieving repeatability is the key to an efficient process that maintains a level of quality in the product. By defining the manufacturing process through the utilization of standardized work, we can control abnormalities and move closer to the ideal manufacturing situation.

It has been said that without standardization, there can be no kaizen. Standardization is so fundamental to the TPS that it literally forms the foundation of the TPS house (Figure 6.11). Without standardization, the TPS is literally without foundation, and therefore would not exist.

Companies will often attempt to implement kaizen without first establishing standardized work. Although some improvement can be achieved,

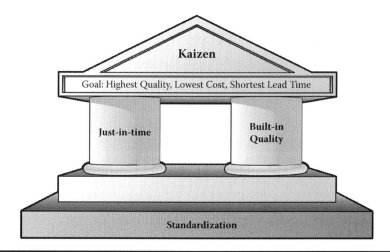

Figure 6.11   TPS House.

it will inevitably not be sustained, as there is no standard in place to reflect the improved process. Therefore, not only is standardized work essential for kaizen, it is also essential for sustainability.

When I am working with a company to improve their manufacturing process, the first step that I have them take is always a simple standardized work exercise. Generally during the implementation process, we will see efficiencies improve from between 10% and 40%. The process of standardization will highlight abnormalities in the process, and once the abnormalities are corrected the process is improved. Although this is a natural process that results in improvement, it should not be confused with actual kaizen, or continuous improvement. Without a formal program of standardized work, improvements that are made in the process are often lost over time. A formal standardized work process ensures that as improvements are made in the process, the standardized work documentation is updated and this preserves the improvement for the future.

To fully understand these concepts, it is important to understand a philosophy that I refer to as the kaizen continuum (Figure 6.12). Simply stated, the kaizen continuum is the path to continuous improvement.

The first step in the kaizen continuum is always standardization. You start with a standardized operation or a standardized task, and then, only then, once you have achieved standardization, can you really make the continuous improvement cycle begin. Once you complete the first cycle of kaizen, then you standardize again and you keep that process going until you achieve the ideal condition. Although the ideal situation is rarely achieved, each cycle that you complete in the continuum theoretically is a step closer to the ideal

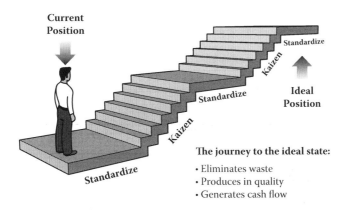

Current
Position

Standardize

Kaizen

Standardize

Kaizen

Standardize

Ideal
Position

The journey to the ideal state:

• Eliminates waste
• Produces in quality
• Generates cash flow

Figure 6.12   Kaizen Continuum.

situation. Therefore, if you follow this process, you can continue to improve the operations. Sometimes people ask me if it is really necessary to take the time to standardize after each cycle of kaizen. My answer is "absolutely." The process of standardization ensures that the improvements achieved in the cycle are sustained. After all, it is easy to create improvement in a process, but the real value is only achieved if the value is sustained and the process does not return to previous methods.

Standardization is also a very practical approach to conducting business. I am often put in the situation where I am introduced into a business and quickly have to ascertain the current situation. I begin with a detailed analysis of the process. All operational processes have some steps of the process that are more crucial and provide more overall value to the final product than others. By understanding the ratio of the two, I am able to understand how much waste is built into the cost structure of the operations. Earlier we discussed this as value-added work (VAW) and non-value-added work (NVAW). The ratio of VAW to NVAW is essential for understanding the cost structure of any operation.

Standardized work is instrumental to understanding the baseline cost structure of the business. For example, if the operation is producing a widget and the widget has five components, it is standardized work that enables the operator to understand the quantity of material, the operational resources, and the labor necessary for producing the widget. Too often, organizations have an elaborate MRP (material requirements planning) or ERP (enterprise resource planning) solution where the process is engineered to produce the component with a specific amount of material, resources, and labor only for the actual execution of the standard to be trusted to an undisciplined process. The operations always have huge variances in material, capacity, and labor.

Standardization is a fundamental criterion for establishing an efficient and effective business process. The goal of standardized work is to develop a process that can be performed repeatedly in a manner that preserves the efficiency of the operation by limiting the unnecessary NVAW, or muda. The process should limit human movement and optimize the utilization of any equipment, tools, and/or jigs.

Standardized work is an effective tool for involving the workers in the creation of value in the process and maintaining standardization in the process. Some organizations focus solely on the equipment and the plant facilities. Their goal is to engineer a process that requires a minimal amount of human involvement. Although automation is a great tool (when implemented correctly) for increasing efficiency, I have seen too many manufacturing processes with a high degree of automation and a low utilization of the workers. This scenario leads to workers who are detached from the process, and both quality and efficiency ultimately suffer.

There is a delicate balance when standardizing a process. I have had many discussions with plant managers, while we are implementing standardized work in the facility, who look at standardized work as a method for forcing workers to complete a prescribed amount of work in the process. Although standardized work is an effective tool for ensuring that a measured level of quality and efficiency are maintained in the process, standardized work should never be looked at as absolute and inflexible. Standardized work should be the basis for improvement in the process. If we again refer to the kaizen continuum, we can see that prior to any kaizen, standardization must be present. The most effective method for determining standardized work is to get the operators involved in the process. Managers tend to think that if you give workers a free rein with determining their own process, they will prescribe a process that minimizes the actual work content in the process. This could not be farther from the truth.

I remember when we were doing a kaizen activity on one of the assembly lines at Toyota's facility in Georgetown, Kentucky. Our demand was increasing for the Camry, and we needed to increase the output on the line by lowering our takt time. We asked each of the production employees to work together with other employees on the line to re-balance the work on the line and to tell us how many processes we needed to add to increase the overall output. It was surprising to find that the employees on the shop floor had designed their processes to be much more efficient than the industrial engineers thought was possible. Management was so concerned that the new processes were too efficient that we actually hired some temporary employees during the initial period to make sure that if there was a problem on a process, we would have sufficient staffing. That turned out to be a big waste of money, as the processes that the workers had developed worked wonderfully. The key to this success was that the workers felt accountable to make the process work because they had an instrumental role in developing it.

When standardized work is developed correctly, the workers' movements are limited to maximize the value created in the process, thus minimizing waste. Standardized work also ensures that only the products that are necessary are produced. By limiting overproduction, standardized work reduces working capital to only what is required. Because the process is standardized, this limits fluctuations in inventory levels.

By only producing the parts that are necessary, standardized work regulates the work and limits the opportunity for defective products to be manufactured. If defective products are manufactured, the standardized work enables the organization to effectively and efficiently countermeasure the process that produced the defective product.

Standardization helps when problems occur in the process. Standardization enables the judgment of normal and abnormal operations in order to detect problems during the process. When the workplace is not standardized, the conditions are continually changing, virtually making it impossible to judge normal and abnormal conditions. Without standardization, the operation is chaotic and unorganized, which breeds inefficiency and higher costs.

6.2.3 Specialize

The third S of the three S's is specialize. Specialization is an important element for the TPS because it is a tool that enables the organization to clarify the roles and responsibilities. Organizations define themselves based on how they measure up to competition. Specializing is an area that can offer a competitive advantage.

I have had the opportunity to examine many organizations, and one of the common traits of a well-run organization is that the roles and responsibilities are defined and understood by everyone in the organization. The absolute opposite is true of failed organizations. Over the last few years that I have worked in private equity, my focus has been working with businesses that need to be "turned around." In a "turnaround" environment, the current course of the company is deemed as not viable and it is up to the person

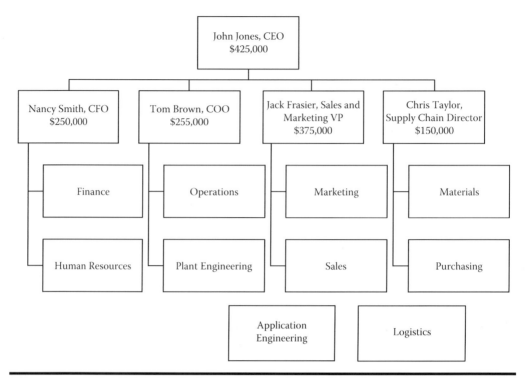

Figure 6.13 Organization Chart.

leading the turnaround to develop a new plan. In these instances, most of the companies that I experience need significant improvement to maintain a level of stability. Many problems exist in an organization that is under stress; one of the most formidable problems in stressed organizations is the lack of defined roles and responsibilities.

One of the first things that do when I go into an organization is to create a functional organization chart with the key members of the management (Figure 6.13). I include the salaries on the organization chart so that I can understand where the cost is in relation to the scope of the work and the responsibility in the organization. It is surprising to see the number of organizations that do not have a simple organization chart. If the organization does have an up-to-date organization chart, it is interesting to see what positions and what functions report directly to the CEO. In family-run organizations, it is typical to see most of the functions report directly to the CEO. Corporate orphans (isolated business units of large corporations) tend to have a good basic structure but are generally filled with too many levels. Distressed organizations usually lack clarity around the roles and responsibilities in the organization. In Figure 6.13, the organization is fairly simple; however, there are two functions that tend to have no direct reporting structure. Neither application engineering nor logistics is formally reporting to any one part of the organization. This lack of clarity usually results in inefficiency in those parts of the organization. People ask, "What is the best structure for a business?" Although there are some basic organizational templates that can be used when analyzing a new organization, there is no cookie-cutter approach that can be used for all organizations. Each organization has its own attributes and requirements that must be understood. The key point is to make sure that there is functional clarity in the organization. Without functional clarity, it is difficult for the middle layers of the management team to understand the direction for the organization.

In an organization where the roles are clearly defined based upon functional responsibilities, the ability to specialize the skill sets in the organization based on the skills necessary in each function should be based upon the knowledge and abilities that are necessary to adequately perform the function. Once these skills have been defined, the roles relating to this area can be specialized to bring like operations together to maximize the value of these skills. An example of this can be seen in the basic structure of an automobile plant (Figure 6.14). At Toyota's plant in Georgetown, Kentucky, the plant is configured physically and organizationally based on the concept of specialization.

Although this example of an automobile plant layout seems basic, it can be used to understand the concept of specialization when it is applied to the roles and responsibilities in the organization. For example, when we consider the workers who are in the trim department, the skills necessary to do their job are quite different than the skills necessary for the workers in the chassis and final

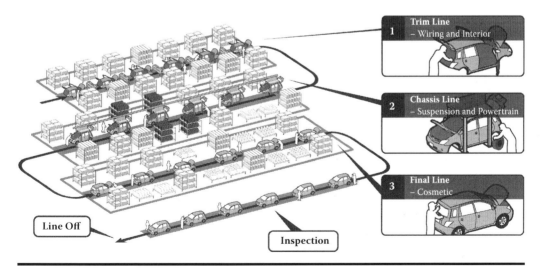

Figure 6.14   Assembly Plant Layout.

department. This is not unique to Toyota and can be found in manufacturing plants with no knowledge of the TPS. As discussed earlier, when looking at the general philosophy of the TPS, this concept makes common sense. Conversely, even though this concept makes common sense, I have been in sophisticated manufacturing companies that did not apply this concept at all. Specializing the work ensures that the best resources are available to perform the necessary work. When like work is brought together organizationally, the training time can be reduced and the quality and efficiency are increased.

Specialization can be applied to all areas of the organization. Specialization ensures that the needs of the organization are lined up with the capabilities of the employees. This process enables the organization to maximize efficiency by ensuring that its resources are contributing to the business by directing the energy of the individuals to the specific needs of the business.

Although these concepts seem simplistic, they are the basis for one of the most efficient and effective manufacturing methods ever conceived. It makes sense to simplify the organization whenever and wherever possible. It makes sense that to produce a product with high quality and a high level of efficiency, we have to have a standardized approach. It makes sense that to get the most out of our organization, processes, and equipment, we should specialize the organization to concentrate the knowledge of the organization.

But as we think about Toyota and we think about the great success they have had over the past thirty years, how did a small company that started out making looms end up being the number one automaker in the world fifty years later? Although there are various business strategies that can be attributed to the success of Toyota, the basis of their success comes directly from the implementation and adherence to the basic fundamental principles that we have discussed as the TPS.

When we look at Toyota, there is no denying that it is a well-run organization. Even if we were to say that Toyota is the premier manufacturing organization in the world, this could be defensible given its recent track record of massive recalls and public quality problems. When we consider Toyota's organizational strategy, we see the commonsense approach that exists in all corners of the organization. Toyota does not have the patent on common sense, and there are many organizations that do a lot of things well. Even though other organizations may not understand the principles of the TPS, common business sense prevails, and many of the successful attributes of these organizations can be seen in the basic principles that we have reviewed.

In my current role working in the world of private equity, I am continually challenged as we look at all types of organizations. When assessing a company,

I determine the best method for applying the TPS in a way that generates value in the organization. Because all organizations are different, this is often challenging. If we remember that the goal of the TPS is to determine the best way to manufacture your products, we can take this one step further and say that when applying the TPS to differing companies, the object is to determine what needs to be done to make the company the best at doing what it does.

Certainly we can use the tools that we have learned from the TPS and apply them in various circumstances to determine the best way. Although I have attempted to simplify this concept, I am not saying that this process in easy to implement. When I was in charge of the cutting-edge kaizen team at Toyota, the hardest part of the job was not to figure out what needed to be accomplished. That was the easy part. The hardest part was to find the simplest, low-cost method of implementation. Unfortunately, there is not a mysterious single element of the TPS that we can take and just apply to any situation and have success. The process for finding success is to look at each situation, determine the best way for that organization to operate, identify what methods will enable them to move closer to the ideal state, and then work to implement that method throughout the organization.

These same concepts apply to all areas of the organization from the executive team all the way down to the front-line hourly employees. When I meet with CEOs, COOs, and CFOs, they usually think that these concepts are great; however, they consider these good tools for the front-line employees and not tools that they need to be able to master. Even once we overcome the senior management team barrier, sometimes by replacing the management team, management think their organizations unique and are too complex for such simple tools to find application. I have heard it all; "Our process is too scientifically precise for this to apply," or "We have a lot of SKUs and these concepts are great for high volume production but not short run multiple SKU operations." These statements just reinforce the lack of understanding that the management team has in regard to understanding the basic fundamental principles of the TPS. I have yet to find a situation where the principles of the TPS are not applicable. It could be said that the TPS is industry agnostic.

6.3 Capital Investment Guidelines

One of the many barriers that I run into when trying to explain to the senior operations person in the organization the concepts of improving efficiency in the operation is the reliance on capital investment to solve all problems. No matter where the discussion begins, in the end, the plans that senior

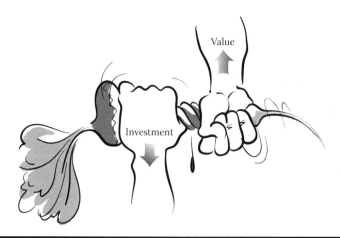

Figure 6.15   Investment and Value Image.

operators have for operations always include capital investment to purchase better, faster machines. It is absolutely true that capital investment can help improve productivity; however, this should always be the last option.

The ability to manage capital investment is one of the most pressing issues I see in the operational realm of organizations. Often the person in charge of operations does not understand how to make improvements in efficiency without adding capital. This quest for capital leads many organizations to the point where the whole organization suffers (Figure 6.15). We once looked at a company that prided itself on having the largest machines in the world. If there is only one thing that you get from this book in reference to capital investment, it should be that you never, ever, want to be known for having the biggest, fastest, or newest anything. In this situation, the purchase of the largest machines caused the company to overleverage itself to the point that the bank took over and forced the family that owned the business to liquidate the company. A modest approach to capital investment is the best bet for long-term financial viability.

Traditionally, Toyota has always had a modest approach to investment; recently, however, it has become evident that the current management overinvested and this led to Toyota's first operating loss in over fifty years. Is this a chink in the proverbial armor of Toyota? I won't go that far, but I will say this is something that will be addressed, and it is unlikely to be a process that will be repeated.

If you have visited many automotive manufacturing facilities, you would be overwhelmed with the lack of automation in a Toyota factory. I have had the opportunity to visit many facilities, and I have found that many of the automotive companies with the worst financial performance are the companies with the highest level of automation in their process.

Although many companies see capital investment as the means for achieving the highest degree of efficiency, this is not always the case. The philosophy in Toyota was always to utilize the capital investment to enhance the manufacturing process with a focus on eliminating work that is burdensome to the operator. It is rare in Toyota that capital is spent solely for the purpose of performing a basic manufacturing function faster.

This same concept can be seen in many other organizations that have adopted similar views for deploying capital investment. I have had the opportunity to visit other auto manufacturers, and both Nissan and Honda also subscribe to the view of minimizing capital investment. All of these manufacturers produce high-quality products very efficiently; however, they do it with minimum capital investment.

Although we can see the frugal roots in the philosophy of the Asian transplant manufacturers, conversely, we can see a completely different approach when we look at the German original equipment manufacturers (OEMs). The German OEMs tend to look at manufacturing from a perspective that is based on engineering. Although this is a much different approach than that of the Asian manufacturers, it is not necessarily a bad thing. Traditional German auto factories are filled with capital equipment that ensures the highest degree of precision. Germans interpret this precise process as high quality. I have a lot of German friends who happen to be engineers. One night we were having a discussion about the difference between perceived quality and actual quality. My German colleagues believed that the more precise the process could be engineered, the better the quality will be on the finished product. It is hard to argue against the concept that a better process produces a better product. My point to them was that because the customer defines what is expected for the perception of quality, any process that produces a product better than the expectation of the customer is wasteful. For example, why produce a car body to tolerances less than 3 mm when it is impossible to detect the difference once the product is fully assembled? In this way, the Germans are spending hundreds of millions of dollars to overengineer their vehicles only for the customers to never appreciate this level of engineering.

When we look at these different organizations, we can say that many of these companies perform successfully. The key point is that capital investment does not necessarily guarantee success or failure. It could be argued that the German OEMs are the only real financially capable contenders to the Asian OEMs. My only word of caution to the German OEMs would be that

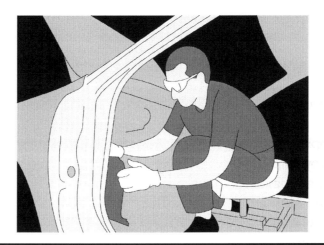

Figure 6.16 Raku Seat Example.

the degree of capitalization that they invest in their facilities burdens them with more financial risk, and therefore the margin of error in the expanding markets becomes riskier. This is really the key with capital investment. When capital can be invested and that investment returns the capital as intended, there is not an issue. The issue is that this rarely happens.

The other restricting point of overcapitalization is the loss of flexibility. It is very challenging to develop a piece of equipment that can be used for multiple products on the same production line. This is why Toyota's philosophy is that people are able to adapt a lot better than machines; therefore keeping people in the process gives the production line more flexibility. In Toyota, capital equipment is generally used for tedious burdensome work or where a high degree of accuracy is necessary. An example of this would be a raku seat (Figure 6.16), which is a type of equipment used to improve ergonomics for an operator performing a specific operation. Figure 6.16 shows that the raku seat enables the worker to be in the best position to complete the operations while reducing the burden from sitting or squatting repeatedly.

Toyota does use robots in the manufacturing process; however, they are generally used for lifting heavy objects, such as a battery or a wheel. In these situations, the robot would be more of a pneumatic assist that could be used for multiple vehicles. This would reduce the implementation costs and offer a broader application to the variation found in a production line at Toyota.

In Toyota, there are some basic principles for equipment capitalization. The five principles applied to equipment capitalization are the following:

1. Tangible return on investment period
2. Recycle, redeploy, and reuse

3. Simple
4. Safe
5. Reliable

6.3.1 Tangible Return on Investment Period

The first principle of equipment capitalization is the tangible return on investment period. Toyota has an elaborate but systematic process for proposing, approving, and implementing capital requests known as the *ringi sho* process. The ringi, as it is known, is an 11 × 17 inch piece of paper with all of the necessary documentation for project approval. The approval level depends on the cost of the project. Generally, ringi requests are approved at the plant manager level and are based on an approved capital budget that meets with the corporate reinvestment budget. The ringi is basically a problem-solving document where the countermeasure is the capital investment. The initiator of the ringi will document the current situation and define the need for capital expenditure. The request will detail how the proposed capital spend will meet the need that has been outlined in the document. The ringi also includes a section where a payback, or return on investment, calculation must be completed. In most instances of capital expenditure, investments with less than a twelve-month return are managed at the plant level and capital expenditures with a greater than twelve months of payback are managed by the corporate engineering level. The general guideline is that if the capital investment will pay itself back during the financial period, then it is easily approved. This differs from a traditional twelve-month payback period in that if it takes three months to get the capital implemented, then the payback period must occur during the financial period. Therefore, in this case the investment would have to be paid back within the remaining nine months of the financial period. This is a great way of managing the engineering resources. Often the engineering department will not allocate their resources and projects will become delayed, which delays the savings for the company. By holding them accountable to this tightened period based on the financial cycle, the projects are usually completed on time.

6.3.2 Recycle, Redeploy, and Reuse

The second principle of equipment capitalization is the concept of recycle, redeploy, and reuse. Although this seems like common sense, many organizations do not look for opportunities to redeploy unused assets, and therefore the complete value of the asset cannot be realized. When considering capital requests in Toyota, the originator has to confirm whether the project can be completed with existing capital that can be recycled for the new project. Often what happens is that an existing machine will be stripped of the critical components, and those components will be utilized during the construction of the new piece of equipment. In Toyota this is referred to as the three "R" check. You would be surprised how often your unused assets can be redeployed to other areas of the operation; many times, they are fully depreciated. When I was at Toyota's Georgetown facility, we had just retooled the welding department with a new production line after nearly twenty years with the original line, and we had a lot of robots that were not capable of the precision work necessary for the welding operation. However, we redeployed these robots into other areas of the operation to eliminate repetitive work from the operators, thus improving productivity and reducing costs. The fact that the robots were fully depreciated yet still functional made this an easy decision.

Once it has been determined that an investment is necessary, the originator is responsible to make sure that the capital project conforms to the final three golden rules of equipment capitalization: safe, simple, and reliable. Even though this seems very basic, it is surprising how complex an unrestrained engineer can make the simplest task. Overengineering is one of the most prevalent problems that I observe when visiting manufacturing operations. In Toyota it is the concept of autonomation, which is to say automation with a human touch, that drives the development and implementation of capital investment. This concept is to use equipment to support human beings in doing the work, not equipment doing the work on its own.

I have had the opportunity to visit many auto manufacturing companies, and even though they are faced with the same problems of manufacturing, they all do things a little differently; this is true even inside individual Toyota facilities. It is understandable that if you visited a Toyota facility, a Nissan facility, and a Volkswagen facility, they might each employ different methods of completing the same task. But why would Toyota, the world's benchmark for standardization, employ different methods in their various facilities? Doesn't this sound very un-Toyota?

Let's say we examine a process that is basic and universal to all auto companies, for example, installing the tire and wheel assembly onto a vehicle (Figure 6.17). Why would Toyota facilities employ different methods for such a basic manufacturing task? In Toyota this concept is often

Figure 6.17   Tire and Wheel Installation.

employed to find the best method. Each facility often will engineer a method to perform the same task a different way to understand which method is the optimum one. Once the optimum method has been determined, it then can be standardized.

To fully understand the concept of autonomation, let's consider another universal manufacturing process to the auto industry, windshield installation (Figure 6.18). It might make sense to have a robot install the windshield,

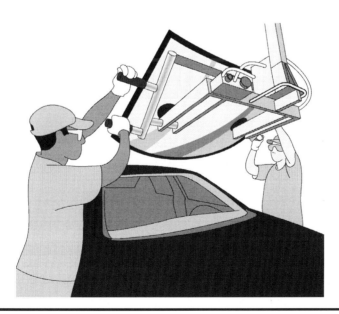

Figure 6.18   Windshield Installation.

since the weight and size of the windshield makes it awkward for the worker to install. To eliminate this burden from the worker, we could use a robot that installs the windshield in the vehicle. Robots seem like a likely fit since a robot is a precision instrument and the task of installing a windshield is a repetitive task. However, there is actually a high degree of variation involved when installing a windshield onto a vehicle.

Variation can cause abnormalities in the operation that can lead to a defective installation. For every problem, there is a countermeasure that can be implemented to solve the problem. Modern engineers have done a wonderful job of developing a host of technological advancements that can be utilized to account for the many variations of manufacturing. Engineers have a large variety of tools at their disposal for solving these types of problems. Whether it is a proximity sensor, a vision system, or another technology, these can easily be added to the robot to confirm the quality of the operation. The problem with this is that the more systems that are integrated, the more complex the operation becomes. If we go back to the original problem of the glass being awkward and heavy, the process has become more complicated than necessary to countermeasure this problem. The process has grown from a robot to eliminate the burden on the worker to a more complex piece of equipment. The more complex the equipment becomes, the more expensive the equipment is to develop, install, and maintain. The more maintenance that is necessary increases the overall costs because more maintenance workers are necessary to maintain and repair the machine. Because maintenance workers are considered skilled workers, they cost more than production workers and this increases the variable cost of the operation.

Toyota strives to go the other way, to simplify the equipment, keeping sight of the original problem. For example, why would you want a robot to install the windshield? You probably wouldn't want a robot to install the windshield for any other reason than that the windshield is a very large component that is awkward to install to the vehicle. And so therefore, in Toyota, most plants use a simple piece of assist equipment to install the windshield. The assist equipment removes the burden from the worker, at the same time allowing the worker to utilize the precision necessary based upon the variables of manufacturing. This is the simplicity of the concept of autonomation, where the human interfaces with the equipment and the equipment is used to either improve quality or to reduce the burden on the employee. In this instance, autonomation takes on the true meaning of automation with a human touch.

6.3.3 Simple, Safe, and Reliable

Realizing that everything that is implemented from an automation standpoint is based on these simple concepts, we can simplify this with three basic principles of mechanical automation:

1. Simple
2. Safe
3. Reliable

From a simplification perspective, the equipment must solve the problem with the lowest level of mechanization. When designing a piece of equipment, equipment that is designed to function mechanically versus equipment designed utilizing electrical or pneumatic components is preferred. Although these systems often require more ingenuity when being designed, they are easier to maintain, less expensive, and more reliable.

From a safe standpoint, we want to make sure that the equipment achieves some basic measure of safety, such as meeting OSHA guidelines. More importantly, the equipment is designed with the principle of placing the least amount of burden on the worker while maintaining simplicity (Figure 6.19). The goal of autonomation is to remove burden from the worker; therefore, the last thing we want to do is implement a piece of equipment that increases the burden on the worker. Often this increase in burden is not intentional, and that is one of the reasons that Toyota spends so much of their engineering resources conducting manufacturing trials.

Figure 6.19   Work Smarter, Not Harder.

Manufacturing trials give the engineers the opportunity to understand the "real-world" application of their designs and to gain important insight from the actual process and the workers.

Finally, when designing equipment, one of the most important elements to consider is reliability. I have seen equipment that costs millions of dollars to install sitting unused because the machine is not reliable. Although it is great to design equipment to be simple, this is not enough when we consider that the uptime target in most Toyota factories is greater than 99.5%. In many organizations that I have had the opportunity to study, the unreliability of the equipment is built in to the efficiency target as a cost of manufacturing. In many of these organizations, equipment uptime generally will run between 80% and 85%. Many companies consider uptime of 90% to be excellent, and others see it as an impossibility.

There are two important factors that have to be considered when considering equipment reliability: detection of operation in delay and TPM (total productive maintenance).

The first factor for maintaining the reliability of equipment greater than 90% is the ability to detect the line stop prior to the line actually stopping. In Toyota, this is referred to as detection of operation in delay. When we can detect that an operation is delayed and we can respond to get the process back into a normal cycle, we can eliminate abnormalities and line stop.

As an example, let's consider a robotic welding cell (Figure 6.20). In this cell, the robot has to complete fourteen welds to complete the cycle. Once the cycle is completed, the part is advanced to the next station. To maintain

Figure 6.20   Robotic Welding Cell Example.

production, the parts have to cycle from station to station every fifty seconds. In many companies that have some type of monitoring system, the process would only produce an alarm if the robot failed to meet the required fifty-second cycle time. If the notification comes at the end of the process, there is no alternative but for the production line to stop, and this will reduce the productivity of the line. Even if the delay occurred at the beginning of the cycle for the equipment, this will not be clearly understood if the equipment's only warning is to stop production.

To improve the reliability of this process, we could track the process of the robot as it completes its process and determine the target for each operation (Figure 6.21). If any one of the operations within the cycle was delayed, the equipment could notify an operator who could correct the operation and maintain the flow of production. The ability to detect an operation in delay is essential for creating reliable operations.

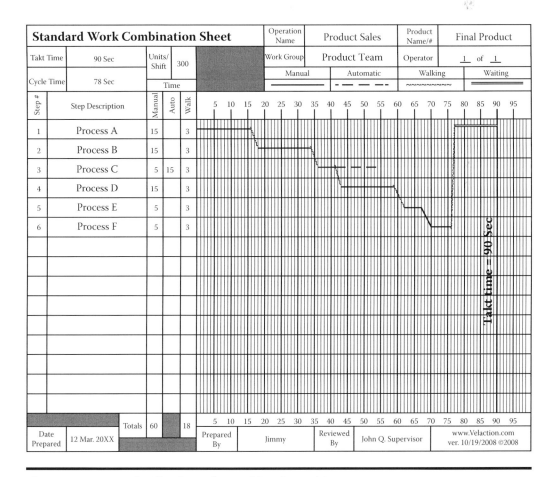

Figure 6.21   Standardized Work Combination Table.

Figure 6.22 Maintenance Example.

The second factor for establishing a reliable operation is to have a process for ensuring the maintenance of the equipment. The most widespread problem in operations with equipment uptime less than 90% is a lack of maintenance of the operation. In many instances, this is a failure on the part of the operational management team to understand the importance of equipment and facility maintenance. I often am puzzled at how easily senior management will spend a million dollars for a new piece of equipment and then will not allocate the necessary operating expenses to maintain the equipment. This is comparable to people who buy a new car and never change the oil or rotate the tires. Proper maintenance is the key to achieving reliability in operations.

The principle of Total Productive Maintenance is that maintaining equipment does not have to be executed solely by a team of specialized, and generally highly paid, workers who understand every facet of the equipment. Even though the majority of the workers are not skilled employees, they can still be involved in the maintenance of the equipment. Involving the employees who use the equipment every day increases the reliability and ultimately improves uptime. Because the production workers interface with the machine continuously, they are more likely to identify abnormalities in the process that could lead to defective operations and downtime.

Even though all of these principles are basic, they should be taken for granted. It is essential for executives and managers who seek to improve their business to truly understand the current situation on the shop floor. Policies and procedures are great, but they are only as good as the organization's ability to implement them. The most gifted and charismatic leader in the world cannot fill all of the gaps in an organization that does not execute at all levels.

Chapter 7

Cost Management
for Profitability

One of the areas that many organizations overlook is the management of operational costs in the business. The one area that the organization can directly control is the cost to operate the business. I have had a great deal of experience looking at the cost structure of businesses, and it is the exception when the management has an understanding of what products actually are contributing to the business. Most managers think they understand, but few actually do. The cost of operating a business is not always properly understood. If a business does not have a complete understanding of the cost structure of the business, the sales organization can be selling products that lose money. The worst nightmare that a CEO can have is a top line that is expanding with products that are not contributing to the profitability of the business. This is exactly what happens when the cost structure of the business is not adequately understood.

7.1 Understanding the "Death Spiral"

Often with a distressed organization, the problem that caused the stress may have originated as a result of external factors. Even though external factors may have put stress on the business, often the leadership of the organization compounds the external factors by focusing the organization on the wrong things. When the senior managers in an organization fail to respond quickly enough to facilitate the turnaround necessary in the business, the company

Figure 7.1   The Death Spiral.

gets caught in a cycle that is difficult to overcome and causes instability in the organization. I refer to this cycle of instability as the "death spiral" (Figure 7.1).

There are several factors that are prerequisites for facilitating the death spiral. First, the company has to face some external challenge, for example, an unexpected increase or decrease in sales volume or rising raw material costs. Next, the company either does not understand their cost structure and how it is affected by the change in these external factors, or their knowledge of the cost structure of the business is not sufficient to produce an adequate response to protect the business from the start of the death spiral.

Most often this external factor is a decline in sales coupled with an incomplete understanding of the cost of the business. When a business gets caught in this cycle, it is difficult for someone in the business to see the problem, and a stressful environment turns into a business in distress. Stress can be healthy for a business; distress is not healthy for any business. Because the death spiral is facilitated by stress plus the lack of understanding of the cost to operate the business, it is relevant to spend some time reviewing the basic principles for understanding and managing the operating costs of the business.

During the recent economic turmoil, most organizations have found themselves with some degree of stress. During this period, many organizations have been managing their finances with a "paycheck-to-paycheck" mentality. The credit markets have been difficult and even nonexistent for certain businesses. This stress on the credit markets affected most businesses by decreasing operating cash, or liquidity. The stress from the credit markets was compounded because the same problem that was affecting the business was affecting their customers and the vendors. Customers were looking for an extension of payment terms, thus increasing the exposure of the business to the market, while at the same time vendors were pushing the organization to tighten terms, as they were faced with the similar pressures from the financial market. Once this process begins, it is very difficult for an organization to recover its liquidity position.

This stress was turned to distress in businesses that did not adequately understand their cost structure. When a business is looking to break the cycle of the death spiral, the first thing that needs to be done before any countermeasure, or kaizen, activity can occur is to understand the cost structure.

7.2 Understanding the Cost Structure

Given this "tightened" market, it is essential for businesses to have an accurate understanding of the cost structure of the business to avoid the death spiral. Many businesses today lack the basic understanding of the cost to do business. Understanding simple concepts such as break-even analysis or cost volume profit analysis can keep organizations from getting caught up in the death spiral. All companies at some time find themselves caught in the middle of a tug of war between their vendors and their customers (Figure 7.2). This is a natural process that managers in the business need to manage during the normal course of doing business. If the margins of the business are healthy, the margin of error for understanding the business is greater and therefore often goes unmanaged. What is surprising, even in healthy organizations, is the mix of products that actually contribute positively to the business and the products that have no contribution to the business whatsoever. I am always amazed once this analysis is completed how the CEO and the head of sales will defend business that does not contribute.

How do businesses get themselves in a situation where they are producing products that do not contribute to the bottom line of the business? This is a good question. Although there are a lot of causes, it all comes back to

Figure 7.2 Tug of War.

the managers in the business understanding the cost structure of the business. It is a basic accounting concept that many businesses make too complex. Understanding these basic principles is essential for an organization to successfully improve the operations of the business.

7.2.1 Fixed and Variable Costs

For management to understand the organization's costs, they must have a complete understanding of the costs of doing business in relation to the volume, or sales level. There are two types of costs; fixed and variable (Figure 7.3). Although this seems elementary, most businesses do not truly understand these costs and the relation to volume, or sales.

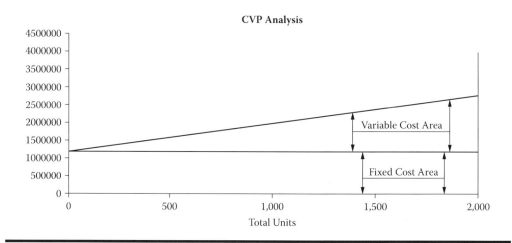

Figure 7.3 Cost Volume Profit (CVP) Analysis.

Fixed costs remain constant no matter how the volume of the business fluctuates, whereas variable costs fluctuate with the level of volume. Fixed costs can be remembered as the "DIRTI 5."

1. Depreciation
2. Interest
3. Repair
4. Taxes
5. Insurance

Other fixed costs are selling and general administrative expenses (SG&A), rent, and others.

Variable costs should fluctuate with the business. If sales volume increases or decreases, variable costs should fluctuate at a level proportionate to the increase or decrease in the business.

Variable costs consist of the following:

1. Direct labor
2. Indirect labor
3. Utilities
4. Supplies
5. Materials
6. Transportation
7. Benefits
8. All other costs

7.2.2 *Minimum Variable Costs*

There are instances when variable costs can act as fixed costs. This happens when sales decline to a level where production is constrained. These costs are referred to as minimum variable costs (Figure 7.4).

Minimum variable costs apply to items where there is not a direct relationship with usage. For example, there is a cost to have basic utility service in the plant whether one machine is running or five machines are running. Even though this condition can exist in several different expense categories, it is important not to confuse these costs with fixed costs. Similar to the concept that we discussed with necessary non-value-added work, minimum variable costs need to be classified correctly in order for the business to determine their cost structure.

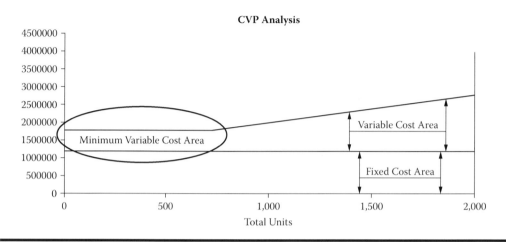

Figure 7.4   Minimum Variable Costs.

7.2.3 Break-Even Point

Adding the total sales line to the cost volume profit analysis, the organization can understand the point at which the organization begins to make or lose money. The point where the sales volume covers the fixed cost and variable cost is referred to as the break-even point (Figure 7.5). Everyone in the business should understand the point at which the business makes money. Even though you can have a highly profitable product when applying standard costing principles, the company can lose money if the total revenue does not eclipse the break-even point. As volume increases above the break-even point, the business produces more variable margin that has

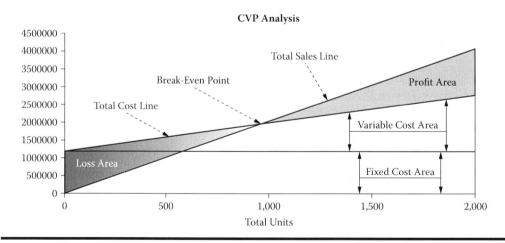

Figure 7.5   Break-Even Point.

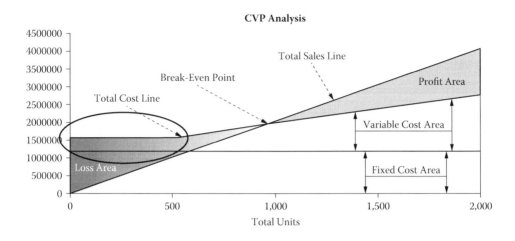

Figure 7.6 Minimum Variable Cost Impact.

a significant contribution on the bottom line of the business. Even products that do not contribute from a standard margin basis can contribute once the break-even point has been eclipsed. This is one of the reasons that CEOs and salespeople will defend negative-margin business.

When the minimum variable cost is considered, the loss of the business is magnified as the sales level decreases. Based on this concept, in a similar method that profits are compounded as total revenue increases above the break-even point, losses are compounded as volume decreases past the break-even point (Figure 7.6). Having a complete understanding of the cost structure of the business is the responsibility not only of the accounting and finance team but also the operations team and executive management.

Reducing fixed costs lowers the break-even point for the business, thus increasing the profit the company can make at lower volume levels (Figure 7.7). Reducing the fixed costs of a business is also referred to as restructuring or, as I like to refer to it, rightsizing the business. One of the first things a business that is experiencing reduced demand must consider is aggressively lowering its fixed costs and minimum variable costs based on the current revenue. This is an easy concept to understand, but many business leaders have a difficult time coming to terms with the fact that the volume may not ever return.

Fixed cost reduction should generally be addressed as a part of the strategic plan, or hoshin, for the organization. Planning improvement to the fixed cost base of an organization is not a new concept and should be done. Rightsizing the business should only be undertaken when the business has sustained a significant loss in revenue where the likelihood of returning to

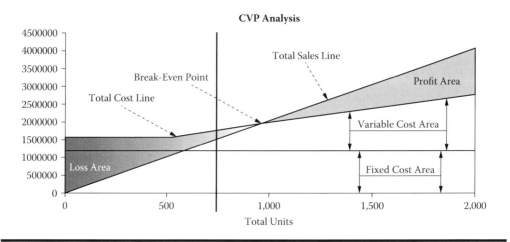

Figure 7.7   Cost Rightsizing.

previous levels is not probable. This occurs when there has been a drastic change in the dynamics of the market. Depending on the magnitude of the change and the relation to the break-even point, the company's level of stress increases. If this stress is not addressed, it can lead to distress in the business. This is the first stage of the death spiral.

Just as decreasing fixed costs increases the profit of a business when sales levels fluctuate, increasing fixed costs decreases the profit of a business, thus increasing the level of sales necessary to cover the cost of doing business, effectively raising the break-even point (Figure 7.8). Increasing fixed costs in an organization should only be investigated when there is sustained revenue growth. Increasing costs should also be a part of the strategic plan, or

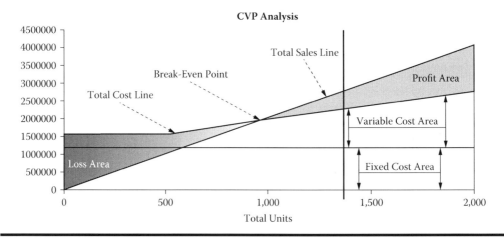

Figure 7.8   Increased Cost Model.

hoshin, for the organization and should not be done in response to short-term fluctuations in the market.

7.2.4 Managing Costs

There are several expense lines on the profit and loss statement (P&L) that impact variable costs. Generally, operating costs are the largest area of variable cost that is directly controllable by the organization. One key to being a successful manager in an organization is knowing what can be controlled and focusing the resources of the business on effecting change in those areas. Too often organizations get off focus and overwhelm the organization with trying to improve the costs that can not be directly controlled by the organization. Examples of operating costs that can be controlled include the following:

- Employee pay
- Number of employees
- Type of employees
  - Salaried/Hourly
  - Direct/Indirect
  - Permanent/Temporary
- Benefits
- Efficiency of the employees
- Scrap
- Overtime

When revenue is stable, managing operating costs is relatively straightforward (Figure 7.9). During these periods, managers must keep costs in line with the budget. Most managers are effective at managing costs when the revenue is in line with the plan. Unfortunately, this is not the normal situation for most businesses.

Most organizations develop operating budgets at the beginning of the year. Management need only to track the expenses in relation to the budget to make sure that the budget is achieved (Figure 7.10). This seems pretty basic. Organizations that have any kind of improvement process will have a plan for some basic level of improvement in the annual budget to reduce the operating costs. To just achieve the status quo, businesses must increase revenue or reduce costs to compensate for annual pay and benefit increases. This activity is planned at the time the budget is created, and resources

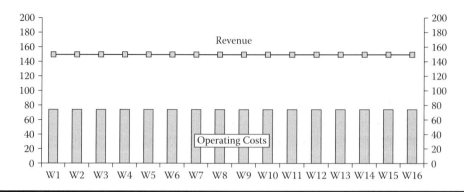

Figure 7.9 Stable Revenue.

are assigned to specific projects to achieve these results. In this situation, managing the cost of the operations is pretty simple, and if this were the case in most organizations, there would be no need for books such as this. However, how often does a year go by when there is no change, positive or negative, to the revenue line?

What is wrong with this picture (Figure 7.11)? Organizations that work to a budget and don't truly understand the impact of the costs of operating the business find themselves trying to explain why margins have been depleted. Often managers get in a mode where they work with blinders on and do not understand the overall impact of costs on the business. The preceding examples are used to explain these basic concepts, and few would disagree with any concepts that have been introduced to this point. It is also important to examine some more realistic examples of what the majority of businesses can expect to see in any given quarter. The next example has been developed based on numerous observations of actual businesses and how the business actually responds to fluctuations in revenue.

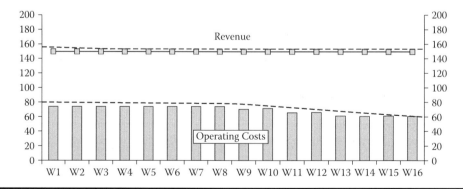

Figure 7.10   Reduced Cost Budget.

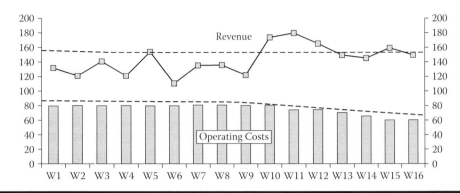

Figure 7.11 Fluctuating Revenue.

7.2.5 *Managing Cost Example*

Figure 7.12 is representative of a typical manufacturing operation. The first thing that is obvious is that sales are below the planned budget by 20%. Operating costs remain at the budgeted level for six weeks before any action is taken. Typically the problem is that organizations do not have real-time methods to monitor the operating costs in relation to revenue, and a correction is made only after the monthly numbers have been reviewed. Once the operating costs begin to be lowered, they are lowered to a level that is not proportional to the loss in sales.

Once these conditions exist, the death spiral begins to gain momentum. Usually the lack of response to a decline in sales is caused by a prominent leader in the organization, who is optimistic that the sales decline is only an anomaly and not a significant event. Often this is the head of sales or even the CEO. At the beginning of the death spiral, the leader convinces the organization that everything will be fine if they can just wait it out. Waiting, or as I like to say, "doing nothing," is the best way to facilitate the death spiral.

In Figure 7.12, the lag in response time was only six weeks; often I have seen the actual condition to be six months or more. Financially this is traumatic to the organization, as the decline in sales reduces the accounts receivable while the costs are incurred at the budgeted level, thus reducing the cash available for managing the business. Even when the company responds immediately, a drop in sales will have immediate effect on the operating capital of the business. When the senior leadership does not have a complete understanding of the concept of the death spiral, they generally make decisions that cause even more stress on the business.

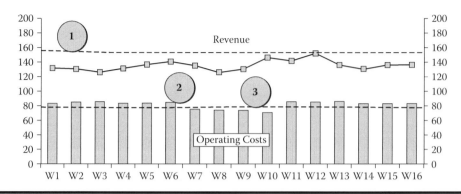

Figure 7.12 Typical Cost Reaction.

Numerous times when I have looked at failing businesses, even when the current ownership is faced with selling the business because they are not willing to fund it further, the senior management has failed to take the necessary steps to break the death spiral.

If we look to Figure 7.13, even though it has taken time for the company to respond to the decline in sales, by week ten the operating costs have been lowered to a level that is in accordance with the sales loss experienced in weeks one through nine. To make things really interesting, we see that sales have increased in week ten. What action makes sense? What is the reaction by most organizations? This is the point where many organizations breathe a sigh of relief, feeling that things are moving in the right direction, and they immediately respond by bringing in resources to handle the increase in demand. The problem with this is obvious. Even though the company took six weeks to respond to a decrease in sales, the response to an increase in sales is immediate. In this example, although sales have increased, they have not risen to the level of the original

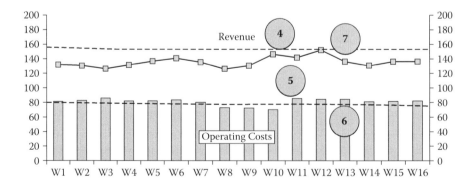

Figure 7.13 Adjusting Cost Example.

budget; however, because the organization has been operating with lower sales and costs at the budgeted level for a period of time, six weeks, when sales increase the natural tendency is to add resources in proportion to the increases in sales, which results in operating costs being higher than the budgeted condition and not in a direct relationship with the increase in sales.

This is a classic example of how management's response, or lack of response, can jeopardize the business. Many companies that are experiencing financial distress do not take the initiative to sort out the company's problems. Many CEOs take a wait-and-see approach, and although patience is a virtue, waiting for the situation to correct itself is certain to produce only one outcome, and that is the death the business. In these situations, organizational leaders develop a restructuring plan for the organization, which paves the road to recovery by planning for increased, or at least recovering, sales. In Chapter 8, I outline a process that takes on the exact opposite approach. The process identifies the opportunities that exist in the current organization within the current stream of revenue and within the current cost base of the business.

Chapter 8

Execution

In this final chapter, I am going to help you put all of the pieces together so that you can execute the principles discussed throughout the book. The basic concepts reviewed here are based on principles taught to me by many teachers during my eighteen years with Toyota. I have taken the principles that they imparted to me and developed them into a system for executing improvement initiatives for the operation of any business. If there is one thing that all companies have in common, it is that at some level all companies operate in a similar manner. Every business buys "stuff" and sells "stuff." From the perspective of this system, the "stuff" is irrelevant. The important thing is to focus on improving things within your scope of responsibility within the operation. If you are a floor super-visor who manages a team of five workers, then you need to be focused on the aspects of those processes that you can control. For mid-level managers who may have the responsibility for a department, you should focus your efforts on the opportunities within your department. There may be many problems passed on from other departments that plague your processes, but if you don't control those operations, you will only become frustrated if you devote all of your energy to trying to solve prob-lems outside of your area of responsibility. If the culture permits, these principles can be used by a cross-functional team of mid-level managers who have the responsibility for the area of concern. If you are a C-level executive, then you have no excuse. If you apply the principles outlined in this chapter, using the tools and methodologies discussed earlier, you will have success. After reading this book if you still have problems you can't fix, give me a call. I am sure something can be arranged ... for a nominal charge.

Organizations that are familiar with the strategic planning process known as Hoshin Kanri are aware of the principle and practice of SWOT analysis. SWOT analysis looks at the Strengths, Weakness, Opportunities, and Threats facing the organization. The Strengths and Weaknesses are internal factors, whereas the Opportunities and Threats are external factors. One of the mistakes C-level managers make in failing organizations is that they have focused on the Opportunities and Threats versus focusing on the Strengths and Weaknesses of the organization. Some senior leaders have a mind-set that looks outside of the organization for the solutions to the problems plaguing it. The danger is that the organization can expend a lot of effort driving toward a perceived opportunity that ends up costing more than the organization is able to afford.

8.1 Facing Reality

I worked with an electronics manufacturing company that was a fairly healthy business with 10% earnings before interest taxes' depreciation and amortization (EBITDA) margins. Wouldn't we all like to have sustained EBITDA margins of at least 10%? The CEO of this business was convinced that the future of the company was to develop business in a foreign market where there was no current market for the company's products. The perceived opportunity was based on the change of a government regulation that would change the dynamics of the market "overnight." In my experience, "overnight" opportunities are nothing more than desperation.

Real sustainable business improvement comes from improving the things that can be controlled within the organization. I do not want to send the message that exploring new markets is bad. Every business strategy has a time and a place. For an organization to expand into a new market, the organization should be experiencing some level of stability, not stress. If sales have declined, the appropriate step is to reduce costs. Once the costs are in line with the reduced level of sales and the business has performed with some degree of stability, then alternative strategies can be reviewed for increasing revenue. This is a difficult reality to face for many business leaders. It is hard for leaders who have worked hard to build a business to admit that the business may not be what it once was. When this happens, the organization needs to respond and make decisions based on the information that is known. When a company experiences this type of stress, the leaders' first priority is to remove the stress. In today's business world, what company is not stressed at some level? Many companies are even experiencing some level of organizational distress.

8.2 The Five-Step Process for Executing Improvement Initiatives

The primary goal of this book is to equip the reader with the information necessary to take the basic concepts of the Toyota Production System and implement them to effect real change within their organization.

In Chapter 7, the basic concepts of the Toyota Production System were reviewed and some examples of how these concepts are applied in real businesses were given. These are basic steps that I have applied in various areas of operations at Toyota and that have been refined to apply to organizations of various sizes, management capabilities, and organizational maturity. I have used this systematic approach to produce results in multi-billion-dollar global automotive manufacturers and in a thirty-million-dollar contract manufacturing operation. In my years of experience, whether it was working with a troubled supplier while with Toyota or with a failing company in the depth of the worst economic crisis we have seen in the United States for more than seventy years, I never found an occasion where this systematic process did not produce significant improvement in any organization. I developed this process to be implemented in a rapid time frame where quantifiable improvement can be realized to the bottom line of the business with minimal capital investment.

The Process
1. Assessment: Understanding the Business
2. Setting the Course—Planning for Change
3. Rapid Implementation
4. Stabilization
5. Continuous Improvement

8.2.1 Assessment: Understanding the Business

The first step for effecting change in the organization is to properly assess the business. Many times when I talk with senior managers who are attempting to implement an improvement process, they have a clear idea of what they would like to achieve and even understand the evolution of where the business should be positioned. One of the most basic principles that managers overlook is that they don't really understand the current situation of the business. Organizational self-awareness is often lacking. Although

Figure 8.1   Kaizen Continuum GPS.

the gap between the current situation and the perceived situation varies from business to business, the one consistent factor is that a gap exists. This is one reason that businesses may begin an improvement process and make some progress only to have their efforts stifled. When this occurs, the people in the organization becomes discouraged and eventually gives up on the improvement process, reverting back to a condition equal to or even worse than the original condition. Trying to effect change without having a thorough understanding of the current state can be compared to trying to plan a trip without knowing where you are (Figure 8.1).

For me, assessing the business not only tells me the current state, it also will let me know the area where I need to begin the improvement process. To effect change in the organization, I want to focus on the area where I am going to benefit the most for the effort exerted. We often call this getting more "bang for the buck" (Figure 8.2). I use the assessment period to lay out my roadmap for change.

It is possible for an organization to assess the current state using internal resources. It is beneficial during the assessment phase to utilize outside resources to provide the organization with an unbiased view of the current state of the business. In an organization with multiple sites, this could be accomplished by utilizing managers from other facilities or by utilizing an outside resource, such as a consultant. Although using an outside resource may be costly, if the resource is capable, then the cost of the assessment will be inconsequential given the potential improvement opportunity.

Figure 8.2 Bang-for-Buck Analogy.

One of the primary tools that I use when assessing any business is the value stream map (Figure 8.3). In the world of manufacturing, this is often referred to as a material and information flow chart. A value stream map can be used to gain a comprehensive view of the business as well as to analyze a process in detail. It is amazing how few organizations actually understand the value stream of their process. Whether it is a manufacturing operation, distribution operation, or sales organization, the majority of organizations fail to recognize the overall benefit of understanding the value stream.

I work in an environment where on any given day I can be called on to work with a wide range of businesses from various business sectors. I have had the opportunity to view hundreds of companies, and I spend a lot of time assessing businesses and understanding how companies operate. Unlike a consultant, I am also responsible to make sure that whatever I think the opportunity is, I am able to initiate a plan that can achieve the level of improvement that I have identified. One of the tools that I utilize to make this assessment is the value stream map.

I have found that most organizations only focus on the part of the value stream that they believe has value, which generally is the core of the business. This could be a plastic injection molding operation that only focuses

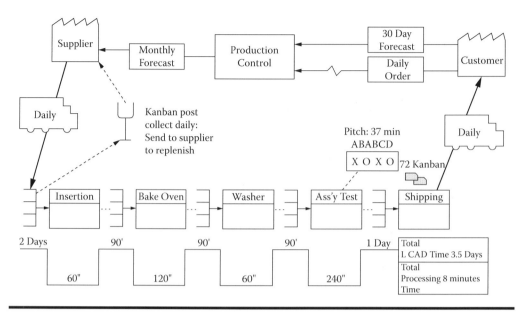

Figure 8.3 Value Stream Map.

on the process to mold the parts and does not consider the other aspects of the operation. This includes basic areas of manufacturing such as conveyance, primary molding operations, secondary processes, assembly, packing, shipping, and inventory management, to name a few.

When we consider that many organizations don't understand the basic manufacturing operations and their contribution to the value stream, then it is not surprising that these organizations don't focus on the broader components of the value stream, such as design, sourcing, procurement, inbound logistics, outbound logistics, customer inventory, and so forth.

This doesn't necessarily mean that a company that does not understand the full value stream of their product or process can't be profitable. Many organizations operate with no knowledge of their value stream and are quite profitable. The fact they are profitable may be the reason that they are not aware of the opportunity that surrounds them.

Most small businesses have limited resources and therefore tend to concentrate their resources on the areas of the business that are critical to preserving customer quality and delivery. An example is an entrepreneur who has an engineering background and starts a manufacturing business that is built around his core competency, engineering. This is very common to see in an entrepreneurial organization.

Because the organization was developed around the technical capabilities of the entrepreneur, other areas that are not related to the technical

Figure 8.4 Mismanagement.

competency are treated as peripheral aspects of the business. For example, supply chain management. When the core competency of the operation is built up as a technical process, the operation has little concern for the supply chain other than ensuring that the raw materials are available to meet the production schedule. In this example, ignoring the supply chain may not have an immediate negative effect on the business; however, the business cannot perform at an optimal level unless there is a complete understanding of the many processes that make up the value stream. The company will only achieve the ideal state by having an accurate understanding of the current state of the business.

Getting the assessment process started can often be one of the most challenging aspects of the process. Members of the operational management team may get defensive when opportunities are exposed. This is completely a natural reaction. Who likes for someone to come into their house and point out all of the things that may be lacking? It is at this stage of the assessment process that I am able to start to assess not only the value stream but also the capability of the management team to effect change.

When I am assessing a business and the members of the management start using phrases like "I understand what you are saying, however we have always done it that way and it is impossible to change," I get concerned for the manager's long-term viability of effecting change in the organization. If a manager makes those types of statements to the owner of the company, how do you think that manager will communicate to the organization?

Another situation is when the management team has the point of view that "if you think this is bad, you should have seen it two years ago." One of the most common barriers to real process improvement is when management attempts to define success based on where they were instead of where they should be. Managers who constantly refer to the previous condition will never meet their potential. To implement a continuous improvement process, it is essential that you are always managing yourself from the ideal situation. It is very easy to improve a business from the current state and then sit back and relax, being satisfied with the progress that you have made. I am not saying that it is not important to recognize progress. I think it is essential to celebrate milestones along the path, as long as the organization stays focused on the goal.

Completing a value stream analysis and a gap analysis is a good start to an assessment process (Figure 8.5). There are several other tools that we have discussed that are useful for understanding the current situation. Break-even analysis and cost volume profit analysis are useful tools during the assessment stage. The key is to gather enough facts to drive actionable change in the organization. This is another reason that it can be beneficial to have another set of eyes look at the situation to help to develop the current state map. An unbiased view is valuable and can bring new insight. Good managers will be open to looking at their organization from different perspectives.

For senior leaders who are assessing their organizations, this is a great time to assess the functional managers in the organization. If you want to implement a continuous improvement process, you have to have managers

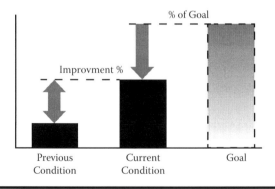

Figure 8.5 Gap Analysis Chart.

who are willing to look at the current state and recognize the opportunity for improvement.

A good assessment should yield the following results:

1. The current state value stream map (VSM)
2. The ideal state value stream map (VSM)
3. The opportunities identified on the current state VSM
4. A project-by-project summary of the opportunities
5. Cost estimates for implementation of the projects
6. Time estimate for implementation
7. Resource requirement
8. Estimation of results (quantified)
9. Return on investment calculation
10. Cash return estimate

In Chapter 4, we briefly discussed the value stream mapping process. This process is essential for identifying the issues that are affecting the business. Once this step is completed, all of that information has to be translated into something useable for the organization. As mentioned earlier, there are a lot of people who can point out the problems. There is value in understanding the areas of improvement, but you can't convert that value into something the business can use without a process for implementation.

During the value stream process, the opportunities that are identified can be looked at as problems that need to be solved. On the value stream map, these problems are indicated as clouds (Figure 8.6). The clouds call attention to the problems. Using the tools we discussed in Chapter 5, we can classify the problems as waste. For the waste that is identified as muda, we can use the waste summary chart, Figure 5.21, to determine the action that should be taken to countermeasure the problem. With the information provided in this book, a business leader can fix 90% of the problems that they encounter. Sure, there are some concepts that we have not discussed. Ask yourself this question: If you could eliminate 90% of your problems in the next twelve weeks, would you jump at the opportunity? The wonderful thing about these concepts is that they produce real improvement from day one.

The easiest and most efficient way to quantify the opportunities is to identify the gap between the current state and the ideal situation. Once the countermeasure has been designed for the process, we can calculate the improvement that we are going to achieve. Seldom is 100% of the waste

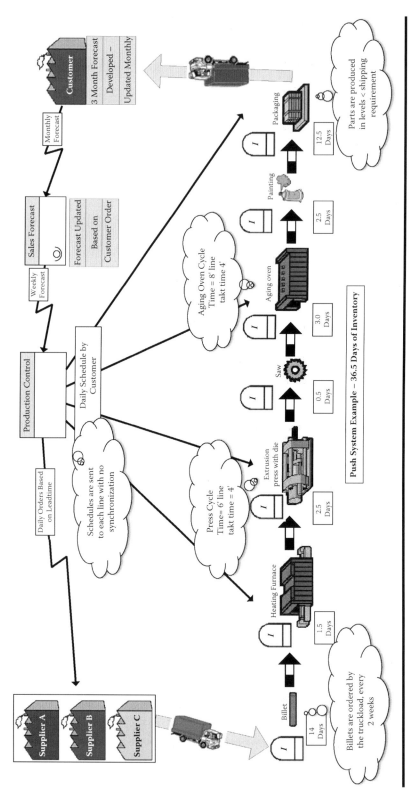

Figure 8.6   Value Stream Map with Clouds.

eliminated; however, by following the process of identifying and classifying the waste, a significant portion of the waste can be eliminated.

When analyzing the opportunities, don't get bogged down with the quantification of the opportunities. It is all right to estimate at this stage. It takes time and experience to identify the countermeasures and to estimate the opportunity accurately. It is tempting to get so focused on the details that it is hard for the company to ever move past the assessment process. I always tell my team, "Don't let the perfect be the enemy of the good." When we are debriefing the project after the implementation is complete, rarely are the countermeasures that are identified in the assessment the exact counter-measures that are implemented. The reason for this is that once you get into a process and get more information, sometimes you have to be flexible and change course. The actual project might change but the overall targets are achieved. Keeping this in mind, you need enough quantification to make an actionable decision.

I like to structure the assessment with a list of the projects, the current state of the relevant key performance indicator (KPI), and the future state of the relevant KPI and then quantify the improvement opportunity. In most operations, a KPI that is always relevant is headcount. Headcount is a good KPI because once the project is over, you can confirm the headcount has been eliminated. If headcount is a KPI that will be used, then equivalent efficiency must also be calculated. This is important because the sales level is destined to change during the implementation, and by calculating the impact in effective efficiency, the project can be assured of an impact to the bottom line.

For the summary of the assessment I am looking for some general details of the projects. This is the step that I warned earlier where it is possible to get bogged down in too many details. At most, each project should have a "before" and "after" slide. Because the projects are meant to be the starting point, or catalyst, for kaizen, providing more information at this stage is just muda. For an example of what type of detail would be in the slides, refer to Figure 2.20 for the "before" and Figure 2.21 for the "after."

Once each project is detailed at a high level, it is important to summarize the costs and understand the return on investment (ROI). To understand the potential benefit for the project, the ROI calculation has to capture all of the implementation costs. This is another area that will only improve with the more projects that you complete. The level of detail should be accurate enough to give you confidence in the ROI, but not so detailed that the proj-ect can never get off the ground.

TPS Project Assessment

Projects Yellow Projects Launched during Boot Camp	Current HC	Future State HC	Opportunity HC	Financial Impact	Current Labor Hours per Unit or fan	Future Labor Hours per Unit or fan	% Improvement in Labor Hours per Unit
1. Project 1	62	47	15	$495,000	5.16 (fan)	4.06 (fan)	21%
2. Project 2	20	16	4	$132,000			
3. Project 3	56	35	21	$693,000	8.86	5.9	33%
4. Project 4	36	30	6	$198,000			
5. Project 5	24	19	5	$165,000	7.5	6.26	16.9%
6. Project 6	6	4	2	$66,000	3.5	2.7	22%
7. Project 7	2	2	0	0			
Total	206	153	53	$1,749,000			

Figure 8.7 Assessment Summary Example.

Projected Implementation Costs

Item	Planned Cost
Implementation Costs	$595,000
Capital & Expenses	$80,000
Estimated Travel Expense	$45,000
Initial Assessment	$35,000
Initial Assessment Travel Expense	$5,000
Total Costs	$760,000
Projected Annualized Achievement	$3,663,000
Projected 2011 Impact	$2,501,000
ROI (Annualized)	4.8X
ROI (20xx)	3.3X

Figure 8.8   Assessment ROI Summary Example.

I am frequently asked what level of ROI I look for in a project. This answer varies from project to project and from company to company. A general rule of thumb for any investment is if the investment will pay for itself in the current fiscal cycle and recover at least one dollar, then that project should be implemented. This is not to say that I have seen a lot of projects that have a one-to-one return. With an operational improvement project, it is rare when the project cannot be implemented and produce an annualized return of three to one. If during the assessment the overall project is close to a one-to-one return, it is necessary to dig deeper into the data to make sure the project really can generate the improvement and that the costs are a little more dialed in to the actual plan.

The other thing to be conscious of is the tyranny of the calendar. If you identify a project for the next fiscal year and you plan to generate one million dollars of savings for the year, the longer you wait in the year to implement, the more actual savings needs to occur. If my project is going to save one million dollars when implemented on an annualized basis and it takes me six months to implement, even if I start day one of the fiscal year, the maximum return I can create for that year is half a million dollars.

Once all of these points are completed and you are confident that you have a good understanding of the current condition and now have a road map for improving the operation, it is time to put together the implementation plan.

8.2.2 Setting the Course

A plan is one thing, but a plan that will execute your project is something else altogether. There are nine key points to developing and executing a successful project plan.

1. Identify the project leader.
2. Completely understand the project assessment.
3. Completely understand the project target.
4. Identify the project resources.
5. Determine the roles and responsibilities.
6. Develop the plan.
7. Completely understand the costs and savings.
8. Communicate.
9. Execute.

8.2.2.1 Identify the Project Leader

When selecting a leader for a continuous improvement project, the leader needs to be technically competent in the principles that are needed for the implementation of the project. It is helpful if the person has technical competence for the operation as well, but it is more important that the leader understand the Toyota Production System principles. If an organization is new to this type of process, I recommend hiring a resource who is technically competent, or bringing in a contract resource.

The project leader has to be a person with the authority to execute the change. For example, if the person is going to implement a project at one plant in a multi-plant organization, the project leader should report directly to the head of operations or the CEO. A project manager should never report to the plant manager of the plant that has to implement the project. This is like having the cat guarding the canary. By creating this direct line to the top of the organization, senior managers can send a message of commitment to the process to the rest of the organization. I recommend this structure even for an organization that has a full-time continuous improvement manager. This position should never report to anyone who does not have the ultimate authority.

The project leader has to have a proven track record for executing projects. There should never be a scenario where a project manager is leading a project for the first time. Sometimes CEOs will get gung ho for implementing lean manufacturing, and they will send someone to TPS classes and create a continuous improvement manager position. This is a big mistake. Although the thought process is admirable, this will inevitably lead to frustration for the continuous improvement manager and the organization. If a proven leader cannot be found in the organization, then one needs to be brought in from outside the organization. If the goal is to have a continuous improvement person inside the organization, that person should work directly with the outside resource. It may be necessary for that person to work through two or three projects with the outside resource before he is capable of leading such an exercise.

This experience of the project leader is very important. Earlier I mentioned that it is rare for a project to implement the exact project from the assessment. For this reason alone, the project leader has to have enough experience that when things go wrong he or she can make the appropriate adjustments to get the project back on plan. The only thing that I can guarantee you about any plan is that it is going to change. Having a leader who can think on his feet is essential.

The project leader has to have great communication skills. This shouldn't be an afterthought; I can't tell you how many times I have seen projects fail because of poor communication.

The project leader has to be a self-starter. Being assertive is necessary for leading any type of project, but especially for a continuous improvement project. Look for the type of person who will push back with senior management in a respectful manner, and once he understands the target, will run through a brick wall to achieve the results. These people are rare, so once they have been identified and have a proven track record of success, make sure they are compensated very well!

8.2.2.2 Completely Understand the Project Assessment

Once you have successfully selected your project leader, it is important that the leader and senior management are on the same page in regard to what is contained in the assessment. It is best if the project manager is identified prior to the assessment and then works as the leader or, at a minimum, participates in the assessment process.

If the leader was not a part of the assessment process, then it is necessary to make sure that there is a complete transfer of knowledge from the resources who conducted the assessment to the project leader.

The leader must have the technical ability to understand all of the necessary tools that will be utilized during the implementation process. The leader needs to go through each project and understand by going to the shop floor exactly what the intention is for each project. If there are any questions or concerns, they need to be understood before the project gets started. If adjustments have to be made, this is the time to make those adjustments. If there is a material change to the project opportunities, then the ROI should be revisited to make sure that the project expectations are understood.

8.2.2.3 Completely Understand the Project Target

It is the responsibility of the senior management team and the project manager to completely understand targets of the project. For the project to be successful, it is the responsibility of the senior management to hold the project manager accountable for achieving the target. If the targets are clear, as we have discussed earlier, there should be no concerns. This is the reason that it is important to understand not only the improvement in the KPI but also the relation of the improvement to fluctuations in the top line of the business. This is why I always calculate effective efficiency in addition to headcount reduction. Using this same example, effective KPIs should be established for all project KPIs. This is very important for achieving success.

It is also the responsibility of senior management and the project leader to be on the same page for the costs of the project. Nothing is more aggravating than for everyone to sign off on the project and the minute the project starts costing any money, the CEO or CFO acts surprised. This is never a good situation.

8.2.2.4 Identify the Project Resources

Now that the project leader has been selected and has a firm understanding of the expectations and the costs for the project, he needs to assemble the resources to complete the project. The project leader must identify the dedicated personnel available to support the project. Projects fail when the resources that are assigned are not capable resources. For this type of process to be successful, the organization needs to put the best resources available in the organization on the project team. This sends a clear message to the organization that the project is important.

I like to use the project teams to evaluate potential leaders for promotion in the organization. When I was with Toyota, I put together a succession planning process that required all high potentials to go through the continuous improvement team.

Once the dedicated resources have been identified and selected, the project team should identify the key resources that need to be involved in the project from the affected areas. This is a good way to get the plant manager of a plant involved in the process. This also sets up the plan for success, as the plant manager will be reporting to the project manager for all matters relating to the project. It is also a good idea to involve the floor supervisors in these roles. Even though they are not dedicated 100% to the team, they can be very beneficial for communicating with the hourly workforce.

Another area of resources that has to be identified is the supporting departments. For a manufacturing process, the supporting areas could include quality, production control, and Human Resources. The more involvement these departments have from the beginning of the project the better the project will flow. Sometimes the strategy is to involve these departments only when necessary; however, the drawback is that having to bring them up to speed once the project is fully developed can bog things down. The benefit of getting the supporting departments involved earlier in the process is that they are more apt to identify potential problems early, and these can be incorporated into the plan. The fewer surprises there are once the plan is in motion, the smoother the implementation process will go.

The project leader should think through each project and identify any maintenance or fabrication support that will be necessary. Because equipment moves and fabrication may take time, these will serve as hard restrictions, or items with a strict time frame, during the planning process and must be considered in advance. It is also the leader's responsibility to make sure that there are sufficient resources for carrying out the plan. In some cases, it may become necessary to bring in some temporary support from other plants for the duration of the project.

Finally, the leader needs to understand the restraints for each resource so that the plan can be properly resourced. The leader must remember these projects usually have to take place during the course of regular production, so it is his responsibility that the impact to current production is minimized.

8.2.2.5 Determine the Roles and Responsibilities

The first key point concerning roles and responsibilities is that everyone must report to the project leader. If the CEO or head of operations is going to continually give direction, then that person needs to be the project leader. This is not a bad idea, and I highly recommend this because then there can be no excuses for failure. If the leader was selected appropriately, then he must have the full authority and responsibility to carry out the project without interference.

The leader should prepare an organization chart for the project organization. The leader should include all of the resources whether they are 100% dedicated to the project or not. The leader should focus on areas of responsibility and not focus on a hierarchal structure. The fewer organizational levels in the project organization, the better. For a project to be successful, it needs one leader and a bunch of workers. When it comes to project management, "many hands make for light work." If the roles and responsibilities are clear and the appropriate resources are allocated, then the plan has a great chance for being successful.

When putting together the project organization, the leader needs to think of the skills and functions that are necessary. Taking a formal step such as this will provide clarity for the team and keep egos in check.

8.2.2.6 Develop the Plan

Finally, it is time to put the plan together. A lot of steps take place prior to making the plan; these steps will ensure that the plan has the best chance for success.

The leader needs to develop the plan with the project team. The team needs to consider any hard restrictions that will affect the project. For example, if a project will only take four weeks to implement but the process is manufacturing a product for a customer with severe quality requirements, just gaining customer approval to make the changes could take longer than the implementation of the project itself. This is especially true when working with the automotive or aerospace industry. These customer requirements need to be built into the project plan. Another restriction that can take time in a plan is fabrication. If there is a lot of fabrication that will have to occur, there will have to be a separate fabrication schedule that will need to coincide with the project plan. Understanding these restrictions is key to developing a good project plan.

Based on the projects that have been identified in the assessment, the project team needs to prioritize the projects. Priorities should be based on the business need and the allocation of resources. For example, if there is a project that is identified that will improve the efficiency of an area by 20% and the current production is already operating at capacity with overtime, this area can be a priority, as it will relieve burden on the plant operations.

The project team also needs to consider the order of the projects. For example, it may be necessary to do projects in feeder lines prior to improving a main line in order to eliminate work stoppages.

When developing the project, it is a good idea to identify key milestones. By identifying these milestones in advance of the project, they can serve as high-level checkpoints to make sure that the project stays on target. The milestones also can keep senior managers from micromanaging the project.

I like to use computer project software to manage my projects. Projects can be shared and managed effectively with a wide array of programs. When using these systems, it is always good to remember the 5W1H rule for making a plan: who, what, where, when, why, and how. Making sure that the project will answer these six basic questions will help you to develop a plan that is clear for everyone to understand and support.

Once the plan has been finalized, all of the project timings have been confirmed, and resources have been assigned, the resource load needs to be addressed. It is the leader's responsibility to make sure that the work is evenly distributed based on each resources allocation. An overloaded resource will become a bottleneck in the plan.

8.2.2.7 Completely Understand the Costs

The project team needs to understand the costs that have been projected for the project. The costs need to be broken down into two buckets: expenses and capital. I like to have the costs broken down into some basic categories for each bucket. For expenses, the costs can be broken down into nine categories:

1. Building and equipment
2. Severance
3. Employee transition assistance
4. Plant inefficiencies

5. Travel
6. Consulting fees
7. Contingencies
8. Permitting/Legal expenses
9. Testing/Quality control

For capital the expenses can be broken down into six categories:

1. Removal of equipment
2. Rearrangement of equipment
3. Inventory storage and movement
4. Equipment installation
5. Fabrication
6. Contingencies

Once the costs have been allocated to these buckets, it is important to forecast the cost by project. This is necessary for individual project leaders to understand the constraints necessary for the project to come in under target. These cost buckets should be scheduled weekly based on the implementation plan. Invoicing dates should be noted, as well as the actual cash disbursements date for the expenses. When working with vendor selection for the project, the terms that the vendor offers should be considered as well as the availability and technical capability. Paying over time allows for a more even distribution of the funds and will enable the company to receive an impact from the savings.

Finally, it is the project leader's responsibility for setting up a system to approve all costs prior to the work being completed. This is essential. The leader should always spend the money wisely, looking for opportunities to reduce costs whenever possible. Even though funds have been allocated for the project, it does not mean that they all have to be spent. It is also wise to plan to come in under the budget so that the project leader has some money reserved in case some items run higher than expected.

8.2.2.8 Communicate

Communication is the key to the success of any plan. Nothing should be assumed when developing the communication plan. It is the project leader's responsibility to develop a communication plan that touches all levels of the organization. The first step of an effective communication plan is to have a meeting with the senior management, plant management, and the project team. Having everyone in one meeting ensures that everyone starts off on the same page. Before the meeting the project leader should develop written expectations for the project and for each individual in the project team. Creating written expectations provides for more clarity and accountability. The written expectations should be reviewed during the kick-off meeting.

For the project team, there should be a weekly report to senior management on the progress of the project. For the weekly report, there should be a template that covers the following aspects:

1. Project KPI
2. Project master schedule
3. Accomplishments from the previous week
4. Activity plan for the next week
5. Cost plan versus actual
6. Savings plan versus actual
7. Fever chart

The format is not important. The report will provide some discipline for the management of the project. One element that I like to include in my projects is what I refer to as a "fever chart." The fever chart is used to rate each member of the plant management team who has a project being implemented in his or her area. The fever chart should also include senior management, including the CEO. Each person is rated as supporting or not supporting the project. I use green to indicate "supporting" and red to indicate "not supporting." It is the project leader's responsibility to complete the fever chart.

I have been through many project milestone meetings where the project leader has reported that everything is on schedule and that everyone is supportive of the project. Later when a problem arises, the leader is quick to point to a member of management as being unsupportive. Using the fever chart enables the project leader to address this as the project is being implemented. If the project is being reported to me, as a senior leader in the organization, I am responsible for holding accountable those who the leader has indicated as "not supporting." This may seem harsh, but it is necessary.

Weekly meetings should be held on the shop floor where the project is being implemented, if possible. As it is not always practical for senior management to be on site every week for the weekly meeting, it is important to establish milestone meetings that correspond to the major milestones outlined in the project plan. These meetings should be on-site and frequent enough to identify concerns and make adjustments so that the project timing can be adhered to.

The project leader needs to work directly with the plant management to communicate to all of the professional staff of the organization. Even though the professional staff may work in an area that is not affected, they need to understand the project and the opportunity that it holds for the organization. This will enable them to understand that there is additional stress on the floor managers and supervisors.

Finally, the project leader and the plant manager need to hold a meeting with all of the shop floor employees. Although it is not necessary to go through all of the details, it is important to give them as much detail as is practical. The key is to be truthful. If the reduction will mean that there are some employees who will be laid off, this has to be shared. The worst thing that management can do is to not be truthful with the shop floor employees regarding the targets of the project.

Sometimes people think that because they work in the office, they are smarter than the workers on the shop floor. This is not the case. The employees on the shop floor are smart enough to know when management is lying to them. Creating an atmosphere where the workforce does not trust management will doom any continuous improvement activity.

8.2.2.9 Execution

If the project leader has followed the steps outlined, the only thing remaining is the execution. As I have mentioned numerous times, plans are just pieces of paper. It is essential for the project manager and all members of the management team to be aligned and understand that it is the organization's project and therefore everyone shares in its execution.

The bottom line is that the failure or success of the project rests with the CEO of the business. If the CEO is engaged and has taken the steps outlined to make sure that the project has the best chance for success, then the project should be successful. If the CEO treats this project as something he can check off his list and come back to six months later, then the project will probably not be successful.

The CEO needs to make sure that the environment is such that the project manager is confident that if help is needed there is some place to turn. If the project manager is experiencing a problem, there is someone who can provide resources to support solving the problem. The project leader should always feel comfortable reporting the true condition of the project and not feel pressured to be overoptimistic during the implementation process.

8.2.3 Rapid Implementation

Now that all of the plans and resources are in place to make our process successful, the plan has to be implemented. Whether this is your first continuous improvement project or the hundredth, the key is to implement the project with urgency. It is difficult to sell the line to the organization that the company is trying to reconcile the cost to the recent downturn in sales and then take twelve months to make the change.

The pace of the implementation needs to be just a little faster than the company's ability to bear the change. This will enable the organization to expand the capabilities for implementing change. When we were planning the first takt time change in fifteen years at the Toyota plant in Georgetown, we took six months to plan and implement the change. Before I left for my next assignment, we were capable of changing takt time with just six weeks of notice. With each successive change that we made, we pushed ourselves further, and through this process we were able to condition the line managers and the workers for this process of change.

Rapid implementation also enables the organization to gain momentum due to the "snowball effect." The same momentum that put the organization into the "death spiral" can be directed toward change and become a powerful tool for the implementation team. As project managers and floor managers have success, they gain confidence; this should be noticed and adjustments made to quicken the pace when possible. There is a delicate balance between urgency and chaos, and it is the role of the project leader to know the organization and be able to increase the momentum when necessary while being able to maintain firm control on the project at all times.

One thing for senior managers and project leaders to watch for is early success that leads to overconfidence. When project and floor managers start taking shortcuts during the implementation process in order to execute the change, this can lead to problems. Keeping the management team disciplined to increase the pace without taking shortcuts is essential for mastering the rapid implementation process.

There are some management tools that I have found to be effective for managing the pace of the rapid implementation process. One of these methods is called the "surge day," or "thrust day." As the project is making progress from an implementation standpoint and the floor managers are working with the workers to stabilize the operation before the next round of changes occurs, organizing an event to focus the organization is a great tool to have at your disposal.

For a "surge day" to be successful, all non-necessary meetings and activities in the plant should be canceled and all of the resources in the organization should be assigned a focus area on the production line. This might entail bringing all of the engineers to the shop floor and identifying pieces of equipment that have not been meeting the new rates. For that day, their job is to stay with the machine and identify what abnormalities occur. Once they have identified the abnormalities, they work to resolve the issue. Everyone should know the goals for the day in advance, and there should be a minor celebration if the goals are achieved. Small things like free sodas from the vending machine are a good reward.

The key is to get the organization to operate at capacity. You will be surprised at how many problems get resolved during these activities. I would even assign roles to members of the senior management team to show our support on the shop floor. This could be a non-skilled job that would show the workers our support for making the organization successful. I always liked to spend the day working in the final assembly area monitoring the level of quality that was being produced. If there were big issues, I would stop the line and bring the problem to the attention of the supervisor for the area and the worker that created the defect. The objective is not to assign blame but to let the supervisor and worker know that the quality produced is important enough that I am willing to spend my day checking for myself. This generally had a positive effect on morale and enabled the organization to stabilize quickly.

8.2.4 Stabilization

As illustrated in what should now be familiar, the kaizen continuum (Figure 8.9), once a continuous improvement project has been implemented, the organization needs to stabilize to fully realize the benefits of the change. Stabilization can often be more difficult than the kaizen project.

Before the kaizen is implemented (Figure 8.10), there may be some areas in the organization that are exposed as problem, or even bottleneck, areas. These areas are the areas that require attention so that the daily operation results in achieving the targets. Because the number of resources is greater

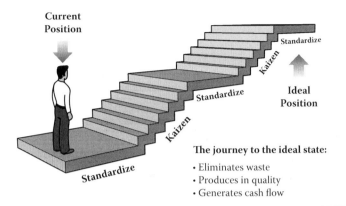

Figure 8.9   Kaizen Continuum.

than the demand from the problem areas, this is recognized as the normal course of business and these areas do not demand immediate attention.

When the kaizen activity is implemented, the water is lowered; the problems that were exposed before are amplified and problems that were being covered up by the inefficiency of the operation are now exposed and causing problems in the operation as well. Even though this seems like an undesirable scenario, this is the desired condition of kaizen. It is only through this process of kaizen that we can "lower the water" and expose our problems. A problem that is not exposed can never be fixed.

A good example of this is inventory levels. I worked on a project in a facility where there were ten days of work in process (WIP) between the prep area and the final assembly area. We were conducting a kaizen class, and one of the teams that was being trained decided to tackle the problem of the excessive inventory. A kanban system was implemented that reduced the WIP from ten days down to six hours. This seemed like a great success,

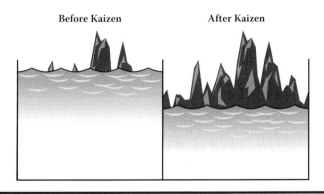

Figure 8.10   Lowering the Water Example.

and from an inventory standpoint it was; however, this now exposed an even bigger problem in the organization. The prep area was not able to maintain the pace and product mix of the final assembly area with only six hours of WIP. Almost immediately, the final assembly area started experiencing downtime. The plant management immediately labeled the project a failure and was ready to go back to the old way.

It is one thing to "lower the water," but we have to be prepared to deal with what we uncover. This is where stabilization comes into play. It is important for all organizations to have stable operations. When an organization is establishing a continuous improvement process, the need for stability is essential for facilitating the cycle.

The shop floor management's discipline for utilizing the basic components of the manufacturing system will be strained during a continuous improvement project. The basic components of the manufacturing system will be the backbone of the stabilization phase. Simple things like tracking production hourly, and supervisors and engineers spending time on the shop floor will make a difference in how quickly the organization can stabilize the operations. There is really no substitute for having a competent management team. There is no process that "runs itself"; for this reason the organization has to be prepared to manage the change.

From a project management perspective, the quicker the operation can absorb the change and stabilize, the sooner the next cycle of kaizen can begin. This needs to be incorporated into the plan and is also something that has to be monitored and adjusted during the implementation process.

Establishing operational KPIs that need to be managed during the implementation of the project is helpful for monitoring the contribution of the current projects. When confirming KPIs, it is essential to monitor a wide range of them. For example, after the implementation, there could be an abnormal level of support provided by the project team for the areas of change. Although it is necessary for the project team to support the changed processes, it is also necessary to manage the level of support that is being utilized to achieve the current level of results. By understanding the current plant efficiency and the level of support provided, the project leader and plant management can determine the appropriate steps for stabilizing the operation.

I have witnessed a lot of really good projects lose momentum and ultimately fail because the project manager and the plant manager failed to make sure the plant was stabilizing before starting the next level of activity. This is the leading reason that many organizations give up on the continuous improvement process. No matter how well the project is managed, there will be some level of disruption to the current condition. If this is managed, the effect to the current operation can be marginalized, but this has to be understood and resourced effectively before the activity moves to the next area or project.

8.2.5 Continuous Improvement

Continuous improvement, or kaizen, is the essence of the Toyota Production System. The kaizen continuum is about creating an environment where the organization is continuously looking for ways to evolve the current process. The process of continuous improvement has to involve everyone in the organization. This is not simply something that can be mandated at the top of the organization and then rolled out like a new 401(k) plan. Continuous improvement is a systematic process that needs to be embraced by the senior management of the organization and then rolled out to every level.

Continuous improvement is not the responsibility of the continuous improvement manager; it is everyone's responsibility. In today's highly competitive global economy, the organizations that can continually improve their processes and products will be the organizations that succeed.

To set the organization on a path for continuous improvement, the leaders must understand that there is a process that needs to be implemented and maintained in order for the organization to gain the results from the kaizen. Too many organizations are looking to skip ahead and try to take shortcuts. The process cannot be cut short; otherwise, the results will not be realized.

8.3 Conclusion

I have layed out a proven system for implementing the basic principles of the Toyota Production System that will drive value in the organization. Some people may see my approach as a shortcut to results; however, I would argue that my systematic methodology for driving the continuous improvement process throughout the organization is a balanced approach. Gone are the days when people could believe that even if the economy tanked, everyone's job would be safe. Ask the workers of Toyota's Georgetown facility. When there is a dramatic shift in the top line of the company, the business must respond in measure, or it dies.

Which Statement is Correct?

TPS is the Best Way.

The Best Way is TPS.

Figure 8.11 Question from a Master.

Any other view in today's economy is not based in reality. Businesses exist in the real world, and for real world businesses to drive continuous improvement, they need an approach that takes the principles of the Toyota Production System and utilizes them to maximize the impact on the organization.

Once the organization has successfully implemented a continuous improvement cycle, it is time for the process to repeat itself. The whole point of the kaizen continuum is that it never ends. There always is the next step that needs to be taken. This is why when management tells me that they have implemented the Toyota Production System, I have to stop and evaluate what they have said. It is not possible to implement TPS. Rather, you are always implementing TPS. This conveys the proper understanding of the system.

Finally, the organization leadership needs to make the continuous improvement process the priority of the company's strategic plan. There are lots of good strategic planning processes, and it really doesn't matter what process the organization utilizes as long as it utilizes a process. Continuous improvement should be how the organization is defined. It should apply to all levels in the organization and should not just be an "oops" thing.

My goal for writing this book is to equip readers with actionable knowledge that they can use to transform their organization into the best organization that it can be. If you only understand one point that I have illustrated, understand that the Toyota System is not the best way of doing something; rather, the best way of doing something is the Toyota Production System (Figure 8.11).

Index

Printed in the United States
by Baker & Taylor Publisher Services